Snake in the Grass

An Everglades Invasion

Larry Perez

Pineapple Press, Inc.
Sarasota, Florida

Dedication

This book is dedicated to my grandmother, Caqui, and my parents, Miriam and Ralph, as an admittedly insufficient token of my gratitude for a lifetime of love, generosity, and encouragement. Together, you colored my formative years awash in childhood riches, and, as a grown man, I continue to strike gold with every passing year. Your selfless support has allowed my wildest dreams, and those of my own young family, to take flight.

Inquiries should be addressed to:
Pineapple Press, Inc.
P.O. Box 3889
Sarasota, Florida 34230

www.pineapplepress.com

Library of Congress Cataloging in Publication Data
Perez, Larry.
Snake in the grass : an Everglades invasion / Larry Perez.
p. cm.
ISBN 978-1-56164-513-8 (pbk.)
1. Snakes--Florida--Everglades National Park. I. Title.
QL666.O6P3525 2012
597.9609759'39--dc23
2011043280

First Edition
10 9 8 7 6 5 4 3 2 1

Design by Shé Hicks
Printed and bound in the USA

Rock python images on cover courtesy of Paul Marcellini, www.PaulMarcellini.com

"*Snake in the Grass* is a gripping and factual account of the most dramatic invasion yet in the continental U.S. It's hard to put down. But it's more than that. Perez depicts the surprising number of invaders in south Florida, the ecosystem-level impacts, and the array of economic forces, politics, and weak regulations that indicate the worst is yet to come."

—Daniel Simberloff, Nancy Gore Hunger Professor, University of Tennessee; editor-in-chief, *Biological Invasions;* senior editor, *Encyclopedia of Biological Invasions*

"Snake-eating alligators and alligator-eating snakes might seem like a science fiction story-line, but those battles are playing out in Everglades National Park, where an invasion of non-native pythons threatens to upend a rich and diverse ecosystem that includes the largest tract of wilderness east of the Rockies. Larry Perez skillfully explores this astonishing assault—the repercussions already tallied and those lying in wait—in a book that reads almost like fiction but which, sadly, is only too factual."

—Kurt Repanshek, founder and editor-in-chief, *National Parks Traveler*

"*Snake in the Grass* is the most comprehensive review of the invasion of Burmese pythons in the Everglades to date, and aptly explains their potential to irreparably alter the dynamics of Everglades ecosystems forever. Burmese pythons, like other nonnative plants and animals, are in Florida to stay. *Snake in the Grass* outlines counter-invasion strategies from legal reform to on-the-ground control techniques. The book could not be more timely."

—Roger L. Hammer, author of *Everglades Wildflowers* and *A Falcon Guide to Everglades National Park & the Surrounding Area*

A narrow Fellow in the Grass
Occasionally rides—
You may have met Him—did you not,
His notice sudden is—

The Grass divides as with a Comb—
A spotted shaft is seen—
And then it closes at your feet
And opens further on—

He likes a Boggy Acre
A Floor too cool for Corn—
Yet when a Boy, and Barefoot—
I more than once at Morn

Have passed, I thought, a Whip lash
Unbraiding in the sun
When stooping to secure it
It wrinkled and was gone—

Several of Nature's People
I know, and they know me—
I feel for them a transport
Of cordiality—

But never met this Fellow
Attended, or alone
Without a tighter breathing
And Zero at the Bone—

Emily Dickinson, "The Snake," 1866

Table of Contents

Color insert following page 126

Preface

The opinions expressed in the following pages are solely my own and do not necessarily reflect those of any governmental agency, organization, or group with which I have previously maintained, or presently maintain, affiliation. Furthermore, unless otherwise noted in the text, my thoughts and conclusions should not be attributed to any of the individuals who have graciously reviewed portions of this work and provided their comments.

This work is intended solely as an in-depth exploration of an important issue and does not necessarily espouse any particular viewpoint or recommend any specific action. Rather, it is hoped the following pages will present a concise account of relevant topics, spark further dialogue among broader audiences, and help advance constructive solutions that benefit both our global community and environment as a whole.

Acknowledgments

There are a great many people to whom I owe a tremendous debt of gratitude in authoring this volume. Even a relatively short work such as this requires the assistance and attention of numerous supporters—usually over a span of years. I am extraordinarily grateful that this project has made me—both personally and indirectly—familiar with the remarkable group of individuals who leave their indelible impressions upon these pages.

I would like to thank Jeremy Conrad, Jeff Carter, Clayton and Ralph DeGayner, Jane Dozier, Dennis Giardina, Cheryl Metzger, Tony Pernas, Joanne Potts, and Skip Snow for generously sharing their time and stories with me. You have all left me in awe of your passion and efforts.

I would like to also express my gratitude to Jessica Demarco, Jim Duquesnel, Alexander Pyron, and Tim Taylor for their generous contributions to the imagery in this work. Special thanks also to Captain Jeff Fobb and Paul Marcellini for kindly giving of their time in helping capture the striking images that grace the front of this book.

I would like to thank those who took time from their busy schedules to review portions of my manuscript and provide their invaluable comments. In alphabetical order they are Mark Davis, Michael Dorcas, Jason Goldberg, Frank Mazzotti, Walter Meshaka, Christina Romagosa, Stephen Secor, and Kristina Serbesoff-King. I would also like to express my heartfelt thanks to both Bob Reed and Tom Lodge for reviewing large portions of this book and providing insightful comments that resulted in a much-improved experience for the reader.

I am indebted also to Richard Grant of the National Park Service for his thoughtful and attentive review of this project.

I also cannot thank enough June Cussen, my editor, and the entire Pineapple Press family for their support of this title and their tireless efforts in the final preparation of this manuscript.

And finally, this work would simply not have been possible were it not for the boundless patience and loving support of my family. In the course of writing this book, I feel I have shortchanged my beautiful wife and two incredible children far too many hours and days. So enough about snakes, kids . . . let's go out and have some fun!

Introduction

Within the confines of a wooden box, most pythons find comfort in collecting their coils and sitting motionless in a corner. So still do they remain that, were it not for the nearly undetectable inflation that accompanies breath, they would appear altogether lifeless. Though strange and slightly disconcerting to us as observers, this knack for deceptive immobility no doubt serves them well in the wild.

Today, however, Damien squirms nervously in his enclosure. Through a pane of Plexiglas I watch as the eight-foot Burmese python elongates his heavy body and slowly probes every crack and crevice of the white-washed box in which he resides. He methodically surveys the walls around him, seemingly intent on finding any available exit. He performs this show for hours—unwittingly providing entertainment for both me and a non-stop gaggle of gawkers who visit my traveling display.

Years of working as a ranger for Everglades National Park have rewarded me with a truly varied career. Working in one of the most dynamic landscapes on the planet offers an endless procession of new challenges, adventures, and duties. Winters usually find me escorting throngs of visitors along narrow trails festooned with hundreds of alligators, birds, and turtles. The arrival of spring brings the possibility of wildfires and a chance to don safety gear and battle fast-moving blazes. Summer spawns foes both big and small—from hordes of ravenous mosquitoes, to the daunting power of tropical cyclones.

Fortunately, this muggy Sunday in April offers a more sedate assignment. Fairchild Tropical Botanic Garden, a popular attraction in southeast Florida, is hosting its inaugural "Everglades Day" and Damien and I are in attendance to provide information about the park to potential visitors. We have been provided a table in a prominent location in the garden, upon which I've thrown a colorful tablecloth emblazoned with the National Park Service logo. On display is a full array of maps, stickers, and literature addressing all manner of present-day issues from restoration to climate change. Behind us I've erected three large banners, custom printed with eye-catching graphics, intended to convey the park story at a glance. All the while, Damien remains in constant motion in his cage at the end of the table.

Before meeting me, Damien was yet another nameless python captured by the park's resource management staff. Now he is being used temporarily for the benefit of public education. I christened him this morning when one of the park biologists dropped him off to me at Fairchild, complete with his sealed enclosure. "I don't think you'll need to get in there," he assured me while showcasing the hefty padlock securing the tank, "but if you do, the code is 6-6-6." With that, I could only guess as to what the snake's temperament was at the time of capture.

Throughout the day, in a conference room not far from my display, scientists from Everglades National Park deliver presentations about Florida panthers, prescribed burns, seagrass scarring, American alligators, and hydrology. Between talks, participants stroll leisurely around my area, where various community organizations have also set up shop. As they mill between tables, I notice nearly everyone is compelled to stop and chat with me. Most are gregarious and eager to learn about my display—but not thanks to my colorful exhibits and handouts. Damien, with flicks of his tongue and an occasional flash of his belly, has easily stolen the show.

Burmese pythons like Damien are a hot topic of conversation for the south Florida community, and they're a growing concern for biologists here and elsewhere. As early as 2000, speculation surfaced that a breeding population of alien pythons had become established in the wilds of Everglades National Park. As these snakes commonly grow to lengths of 16 feet in their native range, the mere thought of such large constrictors invading the wilderness is unnerving at best. As years progressed, theory became fact with the discovery of pythons of various ages, egg-bearing females, and nests. Today, pythons are captured by the hundreds annually, and researchers believe that tens of thousands may now be saturating the area. Necropsies performed on recovered pythons reveal their hunger is satisfied by consuming a wide variety of birds and mammals. And beyond south Florida, scientists and policymakers weigh the odds of invasion by similar species elsewhere.

While his brethren exert untold effects on one of the crown jewels of the American national park system, Damien continues luring the attention of adults and children alike. His charisma is, of course, far from isolated. Everyone loves to *talk* about snakes and, after years of public presentations, I know everyone has a story to tell. That's not to say that everyone *loves* snakes, or that stories of encounters always have a happy ending. But for whatever reason, people share a universal fascination for reptiles like Damien. The history of the Sunshine State is rife with examples of how people have tried to capitalize on our interest in snakes—using them as stage props and marketing gimmicks for tourist attractions, zoos, wildlife shows, and the like. And they've been successful, largely because few of us are truly ambivalent about reptiles. Love them or hate them, an attraction exists that I would never profess to understand.

Perhaps it is this attraction, though, that manifests itself as concern on this day. Throughout the event, several people inquire as to when Damien ate last. "I'm not sure," comes my reply. "We captured him only last week in the park." Conversation then shifts to inquiries about when I expect to feed him in the future. "Never," comes my admittedly heartless reply. My guests seem puzzled at first, but comprehension quickly sets in. Presumably just to be certain, they then ask me what I intend to do with Damien.

There are some things that I would rather not do in the course of my job, but am expected to do nonetheless—and carrying forth such conversations certainly counts as one. After all, I spend the majority of my time extolling the virtues of a treasured landscape, attempting to instill an appreciation for south Florida's cultural and natural heritage. I usually do so in full "park ranger" regalia—green pants, gray shirt, hiking boots, shoulder patch, and gold-plated badge. The flat hat that sits gingerly upon my head casts me as an iconic symbol of environmental stewardship and protection. I am among the ranks of thousands of rangers around the country who work daily to protect some of America's most precious resources. And with any luck, I can encourage others to adopt a similar ethic of conservation.

Yet, the issue of invasive species can sometimes cast us in a different light. The truth, as I share with my visitors that day, is that Damien is simply not long for this world. After our display at Fairchild Tropical Botanic Garden, one of two things may happen to Damien. A select few of the pythons captured from the Everglades are implanted with radio transmitters and again released into the park. These few become living experiments that help scientists learn a great deal about how these animals move and behave in the Glades. They are also instrumental in helping track down additional pythons in the field. To date, of the more than 1,700 pythons that have been captured and documented by park biologists over the last decade, only a handful of snakes have been used for this purpose.

Thus, odds are that Damien will find himself in the company of those much less fortunate. Like the vast majority of pythons recovered alive, he will be euthanized. Once expired, he might then be taken to a nearby lab, stretched across a large, cold table and eviscerated. His limp remains will be used to populate a growing data set that feeds our knowledge of the species and informs management decisions. It is a gruesome, yet necessary, reality—one that seldom sits well with the likes of those presently ogling my display.

And so a familiar round of well-intentioned questions begins. "Couldn't you put him up for adoption, or sell him to someone for a pet?" comes one suggestion. "Why can't they be captured and sent back to their native Asia for release? Aren't they endangered there?" returns another. "With so many running around out there right now, what's one more?" resounds another compassionate plea. Their inquiries are clearly motivated by genuine compassion for the living, sentient being stirring helplessly in its cage, unaware that a day of reckoning draws near.

A poignant testament to his future fate lies before him for all to see: covering the full length of a wooden banquet table, and cascading down either side, I have unfurled the preserved skin of a nearly 15-foot python recovered from the park only a year before. As my visitors sample the rough texture of the mottled brown hide with their fingers, I answer their queries with a rhetorical thought of my own: is it realistic to suppose that thousands of 15-foot pythons could be readily adopted, or transported and transplanted overseas, or rereleased into a landscape we are spending billions of dollars to restore? Any such endeavor would necessarily entail significant risks, considerable costs, or potentially unintended consequences.

What has led us down the path that now mandates the wanton destruction of strikingly beautiful creatures like Damien? Why must organizations and individuals, in benevolent service to our land and resources, now serve as judge, jury, and executioners to thousands of living pythons? And why, despite the size and fearsome reputation of such large snakes, does it wrench the gut, tug on the heart strings, and—for some—seem to generate nothing but bad karma? Indeed, what brings us to this unfortunate crossroads, where every avenue results in a loss?

By two o'clock, the day's scheduled talks have finished. On either side of me, my fellow exhibitors begin to break down their portable exhibits and pack up their materials. I take a cue from them and begin to collect my own goods, packing them away in a neat, methodical manner as I've done so many times before. Though I am certain to find myself at another similar event again in the very near future, I'm equally certain that Damien, my alluring assistant for the day, will not be joining me.

1
Snakes on a Sawgrass Plain

I remember being tired, and a bit bored, by late afternoon on May 22, 2001. It had been a typical spring day at the Ernest Coe Visitor Center, the main contact station in Everglades National Park. During the morning hours, I had spoken with literally hundreds of folks who intended to spend the day touring the Glades by car, foot, and boat. Sporting my usual gray and green uniform, I spent most of the morning greeting the arriving masses, handing out maps and brochures, swearing in Junior Rangers, and answering the most common question in national parks across the country: "Where's the bathroom?"

By midday, as the relentless south Florida sun swaddled the marsh in oppressive warmth, the arrival of visitors became predictably slow and sporadic. By late afternoon, only a trickle of hearty souls ventured in to take advantage of the waning daylight. During such times, I would often keep myself occupied by cleaning exhibits, stocking publications, or, more often than not, reading a book recently purchased from the visitor center gift shop. While I don't recall the details of what I had been doing that day, I remember my boredom palpably giving way to curiosity as a young couple dragged a very large plastic container into the building.

The container itself would have been unremarkable, were it not bound tightly with light rope and perforated by crudely made holes in its top—cues that hinted something was alive inside. Despite well-publicized legal prohibitions on the capture, damage, harassment, or removal of plants and animals found in the park, it is not unusual to have some individuals innocently tote entire plants into the visitor center hoping for a positive identification. Children, in particular, are often blissfully unaware of their transgressions as they innocently pick flowers or scamper to capture the strange lizards and insects they encounter. And visitors have also been known to occasionally bring injured wildlife found in the park to the visitor center, under the erroneous assumption that the park is in the business of rehabilitation. I expected to be greeted by a similar situation as the pair of twenty-somethings made a beeline towards me.

Resting his heavy load awkwardly on my desk, the young man greeted me with a smile and proceeded to relay, in detail, how he and his girlfriend had stumbled upon and captured a large Burmese python earlier in the day near Mahogany Hammock—a popular walking trail located in roughly the dead center of the park. Recognizing that such a discovery was (at the time) a fairly odd occurrence, the pair had stopped at the visitor center before leaving the park to report the incident. It was their intention, they explained, to take the snake into town to an acquaintance in the business of breeding that particular species—no doubt a service to the park from their perspective.

Though the couple was well-mannered and seemingly acting in good faith, a great deal of suspicion immediately, and perhaps unfairly, stirred within me. Years of working in public parks has cultivated in me a healthy distrust that prompts me to be alert for people entering protected areas with pillowcases, nets, snake hooks, or probing questions that could be used to poach plants or animals. Though relatively few in number, there is a pervasive and passionate culture shared by many individuals that derives either personal pleasure, economic profit, or both, from collecting rare wildlife. Whatever the catalyst, orchids, butterflies, and artifacts have historically been pursued with great fervor by enthusiasts. Reptiles have the capacity to evoke in some a similar fanaticism that, for the truly passionate, can overshadow concerns about legality, ethics, and political boundaries. The truly zealous will often risk a great deal in the thrill of the hunt.

In this light, the pair before me had earned two strikes: having proven both their desire and ability to capture large serpents, and conveniently having had the foresight to bring along a large, empty, porous Rubbermaid container with ample rope to bind it. Such gear is not standard fare for a typical outing to a national park—and it seemed evident that the pair was actively hunting reptiles in a park well-known for hosting a diversity of cold-blooded fauna.

Despite my suspicions, I expressed a heartfelt thanks to my would-be volunteers and, without the slightest suggestion of wrongdoing, proceeded to provide a primer on park regulations prohibiting the removal of wildlife. All plants and animals, regardless of classification as either "native" or "exotic," are protected from capture and harassment by park visitors—a necessary stipulation that safeguards critically threatened and endangered species against potential harm resulting from cases of mistaken identity. Permitting a free-for-all on the capture of nonnative constrictors in Everglades National Park could, for example, prove detrimental to the myriad native species for which they are often confused. "I'm really glad you guys caught it, but I can't let you take this animal," I remember saying. The lesson, as I recall, was a tough sell. After some deliberation, the young man reluctantly replied, "You can have the snake but . . . we want to keep our container."

An impromptu search for an appropriate enclosure resulted in a large Rubbermaid vessel of our own. Both containers were placed side by side on the floor. The young woman watched silently as her companion slowly unwound the line, removed the perforated top, and quickly landed a grip around the neck of the thick serpent. Using both hands, he slowly hoisted the snake's massive bulk head-first from one enclosure and deposited him tail-first into the next. Only when the eight-foot snake was coiled tightly in the new container did I bravely offer my services—in near unison, I secured the cover as the young man quickly released his hold on its head.

The transaction now complete, I thanked the couple once again for the service they rendered. Having successfully confiscated the large serpent, I bantered lightheartedly with the pair as we slowly sauntered out to their car in the parking lot. My purpose for the escort was more than mere chivalry—I was eager to note the make, model, and tag number of their vehicle. Shortly thereafter, I would pass this information along to our law enforcement division with a recommendation to be alert for suspicious activity.

Prohibitions that guard against the damage or removal of wildlife in national parks protect nearly all living organisms within their borders—regardless of whether they are plentiful, endangered, native, or from another part of the world altogether. Though such regulations find their authority in governmental policy, they find ultimate justification in our modern understanding of basic ecology. With every new investigation, scientific observation reveals in greater relief how intricately all biological organisms are bound to the living and nonliving agents around them. Each, though clearly discernable from one another, acts as a critical strand in the proverbial web of life we are taught about in grade school. When one strand is pulled, stretched, twisted, or broken, the effects are felt to varying degrees across all its connections. Thus, maintaining the historic integrity of entire landscapes usually requires that each individual component from the largest predator to the smallest piece of wood, be allowed to fulfill its function in its natural place of origin.

Figuring out exactly what constitutes the natural state of a given ecosystem, however, can often prove difficult. Generally, our measure of wildness varies inversely with the proximity and influence of people. That is not to say, of course, that we are not an important cog in earth's biological machinery. Over tens of thousands of years, human cultures have surely lived upon and exerted their influence over nearly every acre of the Americas. Yet only our most recent centuries have brought wholesale change to the landscape. Our relatively newfound ability to harness supplemental sources of energy now permits the total conversion of forests, deserts, and wetlands into permanently managed, artificial environments that cater largely to human comfort and efficiency. The irrevocable conversion of vast swaths of land has been accompanied by a growing realization that urbanization disrupts cycles and processes that directly and indirectly benefit all species—including city dwellers. Today, landscapes devoid of obvious human alteration are typically interpreted as "pristine," and are afforded ever-increasing value for their preservation of natural resources, biodiversity, environmental processes, and recreational pursuits.

As our built environments continue to grow in unprecedented speed and size, greater appreciation is felt for preserving the historic integrity of what wild landscapes remain. It is that same lofty goal that inspired the designation of our first national parks, and continues to drive their management even today. In 1916, the United States Congress formally created the National Park Service. The fledgling agency was to assume stewardship of some of America's most iconic landscapes, including Yellowstone, Yosemite, Crater Lake, and Mount Rainier. As guidance, Congress charged the new agency with a specific mission—preserving landscapes "intact for the benefit and enjoyment of future generations."

Currently, the United States boasts nearly 400 distinct units managed by the National Park Service. They range in size from the 13.2 million-acre Wrangell–St. Elias National Park and Preserve in Alaska to the .02-acre Thaddeus Kosciuszko National Memorial in Philadelphia. Collectively they span more than two dozen management designations, including national seashores, recreation areas, military parks, cemeteries, lakeshores, scenic rivers, historic trails, and battlefields. Without a doubt, however, those that have received the greatest fame, and have etched themselves most clearly upon our consciousness, bear designation as national parks.

Indeed, the Grand Canyon, the redwood forests of the Pacific Northwest, and the Great Smoky Mountains all evoke images of larger-than-life landscapes that boggle logic and common sense with their scope and grandeur. The early years of our national preservation movement focused largely on the attractively textured terrain of the rugged American West, which often afforded vistas of such overwhelming scale they practically begged for protection in perpetuity. Every American, whether living or unborn, deserved a chance to witness the surging froth of Old Faithful, gaze upon the snow-capped peak of Mount Rainier, or marvel at the austere granite face of El Capitan. Such aesthetic marvels are often cited as catalysts in the creation of the national park idea. But in time, priorities would take a radical turn, leaving mere scenic beauty to take a backseat to new considerations of ecology, sustainability, and the preservation of wilderness.

For some, a first glimpse of the celebrated Florida Everglades can prove spectacularly underwhelming. The Everglades, after all, enjoys a worldwide renown similar to that of Yellowstone, Yosemite, and the Grand Canyon. But whereas these popular destinations might greet visitors with picturesque mountain ranges, glimpses of charismatic fauna, or a palette of brightly colored foliage, the Everglades presents itself as a visually monotonous landscape sheathed in drab, muted tones. Unlike the conspicuous herds of bison, deer, or elk that have come to characterize some of our best-known landscapes, mosquitoes are often the most abundant fauna to be encountered in the sticky, hot environs of south Florida. And while parks in the American West boast landscapes so imposing they reduce visitors to insignificant specks of carbon, the Everglades suffers so slight an elevation that reaching its tallest peak requires one only to stand up. From such a vantage point, it is easy to look down one's nose at what some have seen as little more than a miasmic, pestilence-filled swamp worthy only of reclamation.

Roughly 6,000 years ago, the workings of a dynamic climate and rising sea level began to forge the familiar south Florida land mass we recognize today. Over time, this mantle would support colonization by a bevy of plants largely from the south, animals largely from the north and, eventually, people from everywhere. The Everglades, in its present form, is a precocious infant—a youthful ecosystem only several millennia old, yet maturing quickly into an energetic, productive environment capable of surviving the harshest adversity while simultaneously providing generously for its innumerable dependents. To meet these demands, however, the Everglades itself depends upon the most basic of all necessities—a predictably cyclic supply of fresh water.

Fortunately, the extreme southern reaches of peninsular Florida stretch close enough to the Tropic of Cancer to be amenable to this request. While winter months bring mild temperatures and a relative paucity of rainfall, the hot, humid days of summer are routinely punctuated with powerful afternoon thunderstorms that infuse the marsh with water. Every year, the clouds overhead are wrung dry—delivering over five feet of water to the thirsty landscape below. Much of it falls on the open face of massive Lake

Okeechobee—a seemingly boundless natural reservoir whose true capacity is limited sharply by its surprising lack of depth. As the wet season continues to unfurl, unrelenting storms fill both the lake and the Everglades to capacity, spilling waters on a path of least resistance that courses nearly one hundred miles to the south on an overland march to the sea. This lengthy journey is somewhat aided by a steady, though nominal, loss of elevation totaling only around nineteen feet. The waters of the Everglades, though slow and almost imperceptible, are always on the move.

Whereas swamps are usually characterized by the occurrence of stagnant water, the Everglades is truly a river in its hydrology. Unlike the raging rivers that find their origins in the confluence of mountain streams, the waters that wind their way through south Florida's marshes and forests do so at an almost imperceptible rate. In the absence of obstruction and constraint, the sluggish waters do not keep to a single riverbed, but rather, reach out broadly on the landscape to inundate millions of acres along the way. All told, 4,000 square miles are said to have been touched by the waters of the historic Everglades. And though slight, the presence of a detectable current in those waters provided the necessary grounds for author and activist Marjory Stoneman Douglas to christen the system with its other familiar moniker—the River of Grass.

Though the peculiarities of the Everglades may not always bend to our liking, they have—over thousands of years—proven hospitable to an astoundingly diverse collection of life. Early naturalists gave the area high marks for the breadth of rare Caribbean plants found growing along the subtropical coast. The waters of south Florida teem with an assortment of fresh- and saltwater life that has proven central to the people of historic and present-day cultures. Yearly congregations of migratory birds provide spectacular testimony to the nearly 400 species that have been recorded in the Everglades. And joining these is a motley crew of threatened and endangered creatures that have, over time, served to both define the area and infuse the River of Grass with character.

The chance of sighting a manatee, Florida panther, or the rarest of butterflies entices some to spend their lives exploring the wilds of south

Florida. Yet, ironically, the promise of near-certain encounters with wildlife also prompts others to shun the Glades entirely. The Everglades have received legitimate fame for hosting dense congregations of critters of ill repute. They are home to innumerable snakes, including four venomous species. They are the only location in the world where both alligators and crocodiles thrive side by side. And insects can be so copious in the summer, some have said you need to throw a rock through the bugs just to get a decent view.

For some, it is probably fortunate there is no other Everglades in the world. Yet as unremarkable as it might seem, it is perhaps the relative scarcity of places like it that lend it truly remarkable value. Perched delicately between earth's temperate and tropical biomes, the Everglades have amassed an unparalleled wealth of resources worthy of preservation. Recognizing an important opportunity, the United States Congress authorized creation of Everglades National Park in 1935 with the stipulation that the park "be permanently reserved as a wilderness, and no development of the project or plan for the entertainment of visitors shall be undertaken which will interfere with the preservation of the unique flora and fauna and the essential primitive conditions now prevailing in the area."

Since that time, the park has been recognized for the wealth of resources it protects. Within its borders can be found the largest sawgrass prairie in North America, an estuary of national significance, and the largest protected mangrove forest in the Northern Hemisphere. Dozens of rare species are found within the park, and no doubt many more yet await discovery. Quietly and without fanfare, the orchestrated interplay of the living and non-living drives the uninterrupted continuance of vital ecological processes—gas exchange, nutrient cycling, water filtration, erosion control, carbon sequestration, and the like. And though underappreciated as a repository of human culture, the park preserves the ancestral homeland of remarkably advanced civilizations that have long since gone extinct.

Everglades National Park has received global recognition for its inherent values and enjoys protection under three distinct, international accords as a World Heritage Site, Wetland of International Importance, and a Man in the

Biosphere Reserve. True to the intent of the park's enabling legislation, the National Park Service has worked diligently to spare the landscape from further physical modification. And in 1978, roughly 1.3 million acres of the park were designated the Marjory Stoneman Douglas Wilderness Area—affording them the highest level of federal protection currently possible.

Given these efforts, the natural world should find refuge in the confines of Everglades National Park. Here the biological and chemical processes that forge life should continue to run wild and unimpaired. In the absence of synthetic materials, internal combustion engines, impermeable surfaces, chemical pollutants, and artificial lights, native populations of plants and animals should continue to thrive and interact as they have for thousands of years.

Yet despite decades of struggle to protect the Everglades against development, the south Florida wilderness has been dammed, diked, and drained nearly to death. For well over a century, well-intentioned water management schemes have successfully yoked the formerly intractable waters of the Everglades watershed. Today, an intricate system of canals, spillways, pump stations, levees, and wells provides a deceptively dependable supply of fresh water that has fueled the explosive growth of industry and population centers in the southern peninsula. Today, nearly half the original extent of the Everglades has been supplanted by agricultural fields and urban subdivisions, which are afforded flood protection through the deliberate and wasteful redirection of "excess" water to the sea in staggering quantities.

Billions of gallons of fresh water are shunted to the coast annually, plaguing the ecological health and vitality of the estuaries that receive them. And while oysters, shrimp, and shorebirds are forced to contend with unnatural gluts of fresh water, other organisms are condemned to suffer lengthy droughts further downstream. For decades growers, politics, and the occasional tropical storm have governed how and when south Florida's accumulated rains were distributed. At virtually any time of the year, without regard to Mother Nature's historical preferences, the flip of a switch could send a deluge north, south, east, or west. Rarely were south Florida's natural areas the beneficiaries of such technological prowess. Rather, some years, Everglades National Park was denied any inflow of life-giving water whatsoever.

Circumventing the very cycles with which the furred, feathered and photosynthetic denizens had become attuned proved disastrous. Wading birds, no longer privy to the gradual draw-down of water that pooled the prey necessary to nourish their young, saw precipitous drops in their populations. Snail kites and Cape Sable seaside sparrows were driven to near extinction, subjected to artificial flooding that drowned their respective prey and young during consecutive dry seasons. Wildfires, historically hampered in marshlands by long periods of inundation, ran rampant in areas that lay parched for months. And further downstream the estuaries of Florida Bay and the coastal fringe, denied periodic infusions of fresh water from the north, saw pronounced changes in salinity, seagrass, and species composition.

Our efforts, though triumphant in bending the flow of waters to our wishes, showed little mercy towards the creatures that once thrived when the river ran wild. While some advocated boldly for the voiceless denizens of the marsh, they were often outliers among a populace that sought to capitalize on newly available lands. During subsequent decades, however, expressions of anger and compassion would become far more popular. Fueled by the crippling degradation wrought not only upon the Everglades but upon landscapes across the United States, communities began to reconsider the price of economic prosperity. During the 60s and 70s the United States passed some of the most progressive environmental legislation in the world, testimony to the dawning of a new age of consciousness regarding our relationship to the biosphere. This nationwide awakening extended to the furthest reaches of the country—including south Florida.

The last two decades of the twentieth century brought radical changes to our relationship with the River of Grass. Conservation organizations worked feverishly to salvage iconic species from near extinction, including the American crocodile and the Florida panther. Legal battles raged over water quality issues as pollutants streamed into the Everglades from agriculture upstream. Water wars became increasingly commonplace as communities jockeyed for supply during times of drought and fought angrily to keep floodwaters off their interests during the wet season. The south Florida flood protection system, originally built to service a population of around two million residents, was now in service to six million. And through it all, some areas were housing triple-digit population growth in a never-ending sea of suburban sprawl.

The inescapable reality that south Florida was growing unsustainably spurred new discussions on how to steward the area's resources in less damaging ways. State and federal interests began crafting projects and strategies for returning flows of clean water to what remained of the Everglades—in patterns and quantities reminiscent of the historic system. And rather than focusing on individual tracts of land, restoration partners advocated for a holistic overhaul of the entire management system that regulated the Everglades watershed. The broad-scale effort was formalized in 2000 with the authorization of the Comprehensive Everglades Restoration Plan (CERP), a multi-decade, multi-billion-dollar commitment to return America's Everglades to their long-lost glory—while, of course, accommodating the water needs of an ever-burgeoning population. After a century of ruling the Everglades watershed with an iron fist, the time had come to extend an open palm and strike a friendly accord.

Over twenty billion dollars are expected to be invested towards the revitalization of the remnant Everglades. The anticipated completion of CERP will mark nearly a century of conservation efforts—one hundred years of toil in an effort to protect the River of Grass for generations to come. And yet as our community labors to right the wrongs of our past, and curb the disturbance and pollution that has marred portions of a treasured American landscape beyond all recognition, we must be ready to face an unfortunate reality. Even in the midst of restoration, the Everglades is increasingly suffering from biopollution—an ill that grows worse every year and has proven immune to traditional conservation efforts. Though well protected from layers of asphalt and concrete, public lands are seemingly defenseless against the onslaught of foreign organisms that routinely penetrate, populate, and overtake native ecosystems. In the decades to come, should restoration successfully return life-giving waters to the River of Grass, it may be to the benefit of a wholly unrecognizable Everglades.

In looking back at the series of photos taken in 2001 showing me holding the Burmese python I received in the visitor center, I can't help but realize I was smiling in every picture (Figure 1). It's not surprising—I chose to be a park ranger primarily out of a sense of wonder for the natural world in general, and an appreciation for south Florida ecology in particular. To me, Burmese pythons are as worthy of admiration as any alligator, orchid, or egret. I am fascinated by reptiles of every shape and size. While some people feel an affinity for birds or plants, I hold a special place in my heart for anything scaly—particularly snakes. In my mind, any creature capable of climbing so high on the food chain unaided by limbs deserves more than a modicum of respect.

Yet it was also a genuine lack of understanding that allowed me and others to easily dismiss the seriousness of such a seemingly random occurrence. Beginning in the 1970s, several isolated encounters with large constrictors had been recorded in the park. Individual specimens of both the red-tailed boa (*Boa constrictor*) and the ball python (*Python regius*) were captured, as were several reticulated pythons (*Python reticulatus*). As each of the aforementioned species was commonly—and cheaply—available for purchase at local pet stores, it seemed likely that the animals being recovered in the park were releases by overwhelmed owners seeking to relieve themselves of pets for which they could no longer provide care or no longer wanted.

In 1979, a lone Burmese python was found dead along the Tamiami Trail, a two-lane highway that skirts the northern boundary of Everglades National Park. At nearly twelve feet in length, the snake certainly fit the profile of other suspected releases. As is customary in parks and preserves, the observation was recorded as part of the expansive database that provides long-term documentation of the area's natural history. Thereafter, more than 15 years lapsed before another Burmese python was collected.

In 1995, on a chilly mid-December day, a park employee driving along the main park road encountered not one, but *two* Burmese pythons basking on the asphalt. Both individuals were captured in close proximity to one another in an area just north of Flamingo—a remote outpost and visitor facility at the southernmost terminus of the 38-mile-long road that traverses the park. Since the 1980s, park rangers had occasionally reported seeing pythons in this area, but physically collecting two within only ten minutes provided park scientists with considerable fodder for thought. And that one of the serpents was a relatively young juvenile measuring only two feet added to the curiosity of the day's events.

Subsequent years saw a surprising upswing in observations of pythons in the park. Whereas only one python had been collected between 1979 and 1994, eleven pythons were removed from the park between 1995 and 2000. Of these, eight were found in areas near Flamingo—an increasingly evident hotspot of invasion (Figure 2). All but one were longer than four feet, generally considered a length at which Burmese pythons are no longer juveniles. The trend caught the attention of park biologist Bill Loftus and Walter Meshaka, the park's curator and an avid herpetologist. In 2000, the pair penned a herpetological inventory of the park in which they noted that the collection of multiple Burmese pythons of various sizes from a very specific region near Flamingo—coupled with the existence of voucher specimens carefully preserved in the park collection—provided ample evidence to consider the species to be established in the Everglades. Although some criticized their conclusion as being premature, Loftus and Meshaka would ultimately enjoy substantial vindication for their assessment.

Reports and captures of pythons by park staff and visitors continued to increase in ensuing years. Three were captured in 2001, with several more reported. The following year, an unprecedented 14 were removed. Among this number were several juveniles, which seemed to provide stronger evidence of a reproducing population. After all, owners of such small and manageable snakes would have little need to release their captives in the wild.

———

The Anhinga Trail is perhaps one of the most popular visitor areas in Everglades National Park. On a typical winter day, the trail hosts literally thousands of pedestrians hoping to catch a glimpse of everything for which the Everglades is known. The network of deep water canals and lakes in the area remain wet even during the harshest droughts, providing an irresistible draw for the aquatic denizens that call the Glades home. Thousands of fish, snakes, birds, turtles, and alligators crowd the area, providing visitors the opportunity to witness interactions reminiscent of a National Geographic special—and all from the relative safety of an elevated boardwalk. A popular trail spur, known informally as Frankie Point, overlooks a relatively small patch of elevated ground, allowing visitors to feast their eyes on the dozens of alligators that routinely compete for prime real estate.

Visitors to the Anhinga Trail got an unusually exciting spectacle in January of 2003. At first glance, the gaggle of people looking out over Frankie Point might have considered the sight before them rather ordinary for the Everglades—a conspicuously large alligator approached, swimming toward them with a snake held tightly in its jaws. But as the fearsome predator drew near, it was apparent his early-morning snack was rather extraordinary. The snake, still very much alive, was roughly 10–15 feet in length and coiling the remainder of its linear mass tightly around its captor. (Figure 3)

Though appearing somewhat compromised, the alligator managed to lumber onto a nearby patch of dry ground. Now exposed, the details of the duel were easier to discern. Though the alligator had grabbed the python just behind the head in its toothy jowls, the remainder of the snake snugly girdled the trunk of the now static alligator several times around. There the pair remained locked in battle for approximately 24 hours, before a rotating cavalcade of hundreds of wide-eyed visitors and park staff. Little seemed to change during that time, save that the snake seemed to grow increasingly limp with every passing hour. By the morning of the next day, observers had written off the serpent, as it lay motionless and still—no doubt punctured to deflation by the alligator's piercing grin.

When threatened, alligators will sometimes open their mouths and hiss loudly in an impressive defensive display. Opportunities to observe this behavior are plentiful along the Anhinga Trail. Thus, when a slightly larger alligator arrived in the area that morning and pulled close to investigate the scene for itself, it was almost expected that such a display would ensue. That the successful hunter began to open his jaws wide to avert a confrontation was not surprising. What was surprising, however, was the near-instantaneous resurrection of the snake everyone had given up for dead. In the blink of an eye, the serpent tensed its sinewy muscles and darted like a shot into nearby vegetation—never to be seen again. Though speculation ensued about whether or not the python was mortally wounded, this much was known for certain: the snake had tangled for 24 hours with the monarch of the Everglades marsh, and somehow, it had ended in a draw.

Visitors to the trail that day were privy to history—the first time ever that an Old World python had ever been observed tangling with a New World American alligator. The size of the animals, the prolonged nature of the duel, and the ultimate outcome helped stoke healthy media interest. Amidst the ensuing interviews with local reporters, park scientists were left to wonder if anything in the Everglades could successfully kill and consume a fully grown Burmese python.

Exactly where alligators and pythons ranked in the Everglades food chain would remain unclear for nearly two years. A second alligator/python encounter was captured in a series of photographs along the main park road in June of 2005. In the open marsh of Taylor Slough, the alligator was clearly photographed pointing its snout upwards, repeatedly throwing back remnants of a large python like a bucket of raw oysters.

The 2005 encounter, unlike that at the Anhinga Trail, failed to garner any significant media attention—perhaps because it did not seemingly defy the natural order in the River of Grass. Still, it was of significance to the park. Before the encounter recorded in Taylor Slough, it appeared the only organism capable of purging Burmese pythons from the Everglades were people who, like those I encountered in the visitor center, came armed with Rubbermaid containers and questionable motives. Confirmation that at least one alligator had successfully taken its rightful throne as the apex predator in the Everglades was cause for some optimism. Perhaps there was hope that the natural system might self-regulate, that a biological means of control might be found to keep pythons numbers in check. Perhaps there was some reason to suspect the Everglades, having survived a century of unrelenting change, might also be capable of surmounting this latest plague. Unbeknownst to park staff, such guarded optimism was destined to last only four short months.

2
Getting Acquainted

Michael Barron has flown helicopters over the vast horizons of Everglades National Park for years. During that time, he has piloted researchers to every remote corner of the backcountry, scoured the landscape in search of bird rookeries and alligator nests, and ferried countless firefighters to the front lines of sprawling conflagrations. And yet despite all he has seen in the wilds of south Florida, little could compare to what he encountered while hovering over the marsh one day in September of 2005.

Flying high above the northern reaches of the park with a pair of researchers on board, Barron saw an unusually shaped figure below. Deciding to investigate, he set the floats beneath his craft gingerly into the inundated sawgrass prairie. Exiting the vehicle, he sloshed a short distance through hip-deep water to find a gruesome scene. There, floating amidst emergent vegetation and open water were the decapitated remains of a nearly 13-foot Burmese python. The animal was badly bloated, and protruding from its ruptured stomach were the tail and hind quarters of what appeared to be a large alligator. Barron surveyed the scene briefly, but didn't stay long. "I started getting a little nervous about being up to my waist in the water," he later recounted, "in an area where there's twelve-foot pythons hanging around." Barron managed to snap a few quick photos before returning to his airship for the journey home (Figure 4).

The subsequent day, Barron piloted park biologist Skip Snow out to the area to examine the macabre curiosity he had found in the marsh. Upon landing, Snow set to work. Conducting a necropsy in the field, Snow determined the alligator (now in an advanced state of decay) to be over six feet in length. In the python's intestines, he found large pieces of the alligator's skin.

Though the puzzling scene provided little to explain why the serpent perished, how its stomach ruptured, or how it lost its head, one thing seemed clear—at some point the python had managed to subdue and consume the monarch of the Everglades marsh.

Media interest flared yet again as news of the discovery spread. Snow was interviewed extensively, and Barron's crime scene photos were published around the world. Numerous national papers detailed the story, touting banner headlines reading "Clash of the Titans" and "Fatal Indigestion." Internationally, the gory pictures of the eviscerated serpent were bandied about by the BBC, *The Sydney Morning Herald*, and *The Daily Telegraph*. And taking a cue from *America's Most Wanted*, National Geographic aired an hour-long documentary that attempted to expose details behind the fatal encounter. Using sophisticated computer animations and laboratory experiments, the episode only added to the confusion by offering a new hypothesis that implicated a second alligator in the scuffle. Despite extensive investigation, the curious double homicide would forever remain steeped in mystery. But for many viewers and readers, the resulting coverage and publicity provided their first glimpse of the Florida Everglades—one far different than the idyllic picture-perfect sunset over an open expanse of sawgrass.

In addition to providing new insight regarding the mutual relish with which alligators and pythons might consume one another, Barron's gruesome discovery also provided an important new data point. The amazing alligator-eating python was one of the first of his ilk to be found in Shark Slough—the liquid heart of Everglades National Park. Not only were pythons now known to be capable of swallowing large alligators, but it was becoming increasingly apparent that they readily occupied the same watery haunts.

In a coincidence that mirrored a bad publicity stunt, ABC premiered *Invasion*, a suspense series about aliens infiltrating the Florida Everglades, that same month. The show featured sets that barely resembled the Glades, starred impossibly attractive actors playing park rangers, and followed a storyline as thin as spider's silk. In only a year, the show's viewership went south, but in the true Everglades it seemed real invaders were headed farther north.

Historically, the Burmese python (*Python molurus bivittatus*) has long proven difficult to categorize—so much so that successive generations of taxonomists have waffled about its proper classification. For the better part of a century, it has been defined as one of two closely related subspecies that occupy a large swath of south Asia and several associated islands. The Burmese python and Indian python (*Python molurus*) are so closely related, in fact, that some scientists continue to argue their true familial ties. Physical differences between the snakes are minute, and include slight, but fairly consistent, variations in color pattern and scalation. But it is the reproductive preferences of these subspecies in the wild that poses perhaps the greatest argument for elevation of each to specific rank. Though interbreeding yields healthy offspring in captivity, there is an apparent reluctance between subspecies to mate in the wild, despite overlaps in the range of each. The ambiguities in their life histories yield two camps of thought—"splitters" who opt to maintain their sub-specific distinction, and "lumpers" who regard both races as a single species.

When viewed collectively, the species is a habitat generalist, capable of thriving in ecosystems spanning a variety of elevations—from sea level mangrove swamps to lower mountain forests sometimes as high as 3,000 feet. Its natural range extends to areas of both tropical and temperate climes—from semi-arid grasslands and deserts to wetlands of abundant rainfall. Regardless of where they are found, these pythons normally utilize the full gamut of natural features in an area, frequenting burrows, trees, rocky outcroppings, riparian zones, open water, and disturbed lands. In so doing, the snakes exercise their astonishing ability to swim, climb, and contort their bodies to meet the demands of the landscape. They are aided, of course, by tools forged over time through evolution: a supple skeleton, sinewy layers of muscle, and strong prehensile tails.

The species can adapt surprisingly well to hostile environmental conditions. In the northernmost reaches of its range, for instance, pythons are able to endure harsh winters during three-to-four-month periods of hibernation. In areas of constant inundation, the heavy-bodied pythons prove semi-aquatic in nature, comfortably spending significant time maneuvering their bulk both above and below the surface of the water. Pythons have also been known to persist in both disturbed and human environments, though it has been noted by some that they fare best amidst more natural habitats. And though it is also not their preferred haunt, pythons can even tolerate exposure to salt water for short periods of time while they comfortably navigate coastal waters.

Indian pythons and Burmese pythons occupy fairly distinct ranges south of the Himalayas. The former occupies nearly the entire South Asian subcontinent, where it is found primarily across large swaths of Pakistan, India, Nepal, and Sri Lanka. By contrast, Burmese pythons are generally encountered further east along a continuous range from northeastern India to southern China, with smaller, isolated populations also persisting in Nepal and on the islands of Java, Bali, Sumbawa, and Sulawesi. The details of natural distribution and biology are only partly known—informed by a limited collection of observations and studies from the field. But over the years, the study of captive animals has helped shed light upon the life histories of many snakes—particularly the Burmese python.

Hatching from the egg at only twenty inches, Burmese pythons experience one of the fastest growth rates known among snakes. During their first year of life, some can grow half a foot per month—particularly if fueled by regular feedings. Pythons reach sexual maturity sometime between reaching five and eight feet in length, a goal which can take as few as two to three years to achieve. The pace of growth slows gradually with age, but nonetheless permits both sexes to attain double-digit lengths. The females, however, typically dwarf their mates—sometimes reaching the length of a stretch limousine and tipping the scales at several hundred pounds.

To accommodate such growth, Burmese pythons readily consume any meal that satisfies their hunger. Diet is generally restricted to warm-blooded terrestrial animals, leaving a wide variety of potential prey on the menu. The python's prowess at capturing, constricting, and consuming large prey has made it a staple of wildlife documentaries. But in addition to mind-boggling meals of leopards, deer, and antelope, the python feeds upon a much greater spectrum of both large and small fauna that includes wading birds, rodents, porcupines, bats, and domestic livestock. Though mammalian and avian prey is clearly preferred, the snakes are ultimately opportunistic in their feeding—known to consume even the occasional lizard, frog, or toad.

Cryptic coloration and a predilection for slow, deliberate movements provide an effective strategy for pursuing a meal. Burmese pythons are ambush predators that patiently lie in wait for their prey to advance. Engaged in attentive repose, pythons silently gather data using a sophisticated array of serpentine gadgetry—a sensitive inner ear transmits minute vibrations on the ground, a forked tongue processes molecules of odor captured from the air, and a full network of facial pits detects slight changes in temperature nearby. It is unclear whether it is motion, sight, or smell (or some combination thereof) that triggers the snake to feed.

Whatever the cue, it commands near-instantaneous strikes—widening jaws bearing rows of wretchedly recurved teeth capable of ensnaring an unsuspecting passerby. This bite, though no doubt painfully alarming, is nonfatal. Lacking venom, Burmese pythons must subdue their meal by sheer force. Using its mouthful of sharp barbs to hold fast the unfortunate captive, the snake throws coil upon coil around the body of its victim, exerting ever-increasing amounts of intense pressure. So strong is this fatal embrace it can at once squeeze breath from the lungs and render the heart inert. After some time, when movement is no longer detected, the large constrictor slowly relaxes its vicelike grip to finally indulge its appetite.

Though people are fascinated by the size, girth, and predatory instincts of the Burmese python, it is in other respects a fascinating biological creature of inspired design. Their seamless camouflage and capacity for stillness are well-developed traits that serve them not only in ambushing prey, but also constitute an effective strategy for self-preservation. A strong, prehensile tail allows them to hoist, suspend, and cantilever their bulk into trees high above the ground. The series of thermal pits that accents their lips allows for continued monitoring of their surroundings—enabling the detection of even minute differences in temperature nearby. And like other primitive snakes, Burmese pythons boast conspicuous pelvic spurs, which aid them in the requisite tickling and grasping that accompany courtship.

The reproductive habits of several python species are well documented, partly thanks to observations in the field, and partly due to the frequency and ease with which they are bred in captivity. Burmese pythons show great variability in the number of eggs they lay. While most will typically produce several dozen eggs at a time, both single-digit clutches and those in excess of 100 have also been reported. In captivity, females can be induced to lay a clutch every year, but in the wild they are more likely to produce a clutch every two to three years. During the time necessary for gestation and incubation a female python will often forgo all food—a prolonged period of four to six months during which she grows increasingly lean. Once her eggs are laid, a female python will swaddle her unhatched progeny in a protective stack of body coils. Should the ambient air begin to cool around them, the vigilant mother will begin to repeatedly contract the muscles in her lengthy body in a rhythmic fashion. As temperatures drop, these full-body spasms will grow ever more rapid and her coils will grow taller and tighter in a bid to generate heat for her unborn young. It is a display of maternal dedication rarely found among other reptiles, and a feat relatively unknown among cold-blooded animals—for a short span of time, she is actually capable of regulating her own body temperature for the benefit of her offspring.

Many facets of the life history of the Burmese python remain poorly understood. The extent of their distribution in certain portions of their native range still remains uncertain and in need of further study. The social workings of these large creatures remain somewhat shrouded in mystery, though observations of captive pythons provides some evidence to suggest a clear hierarchy may exist—particularly with regards to courtship. Both the density of Burmese python populations and the relative impact of disease and parasites on them still require further investigation. Even the full reproductive mechanisms of this species remain in question—some evidence exists that females may be capable of producing clutches of genetically identical young asexually through a process known as parthenogenesis.

The longevity of Burmese pythons in the wild also remains a great unknown. What little can be speculated about their span of life can only be gleaned through records of captives, some of which suggest exceptional animals can reach ages in excess of thirty years. In fact, one of the oldest known captive Burmese pythons died in 2009 at the ripe old age of 43. At the time of her passing, Julius Squeezer measured 18 feet long and weighed 220 pounds. Marty Bone, an avid snake enthusiast, had shared his home with Julius for 35 years, having acquired her as a full-grown adult. Bone attributed her exceptional longevity to both the freedom she enjoyed and the affection he showered upon her. Bone allowed Julius unfettered access to his home and, over time, she reportedly learned how to open doors merely by wrapping her body around doorknobs. Marty modified furniture in his home to better accommodate Julius, and even allowed her to curl up with him in bed. "At night she'd lay her head on me," Bone recalled. "She was my bedmate, housemate . . . she was special to me."

Burmese pythons are strictly an Old World species. Still, much of our current knowledge about their natural history has come squarely from our experiences with them in the Western Hemisphere. Zoos, dealers, and private collectors have provided detailed information on their reproductive habits. University studies have furnished information on social behavior, feeding response, energy efficiency, and the effects of visual deprivation. Even the sad passing of Julius Squeezer in a private Utah home provided an important bit of data about potential species longevity.

The establishment of Burmese pythons in south Florida provided a simultaneous need, and opportunity, for further study and data collection. By 2006 it seemed evident that the python population in south Florida was growing larger. Just a few years prior, park employees were removing only two or three of them from the area annually (Figure 5). But beginning in 2002 the yearly figures began to make surprising jumps—staff removed 14 individuals in that year alone. The following year they removed 23 more. In 2004, the count jumped yet again to 70, then to 94 the following year. The park seemed poised to hit triple digits by 2006—a sobering milestone considering it only reflected those snakes that were captured and removed. The annual tallies didn't include the many animals reported that subsequently got away. Nor did they reflect those that remained altogether unseen.

Increased awareness of the issue, the hiring of additional staff, a growing proficiency for capture, and a more concerted communications effort were all no doubt complicit in yielding higher rates of removal with every passing year. The clear statistical trend, however, was still unsettling. Evidence continued to mount suggesting the snakes were freely breeding in the wild. Park staff had thus far found pythons of various size classes throughout the Everglades, including juveniles. Newborn pythons had been recovered, some still bearing the ephemeral umbilical scars where their newly shed yolk sacs were once tethered. And several gravid females had been recovered—their innards packed to the walls with fertile eggs (Figure 6). Though the evidence of breeding was certainly compelling, it was not technically conclusive. Scientists had not yet happened upon the necessary "smoking gun"—neither copulation nor a nest site had been documented in the wilds of south Florida. To do so, scientists would need to learn as much as possible about the habits of this Old World serpent in its New World haunts.

Exactly who is to blame for the introduction of Burmese pythons into the Florida Everglades is a topic of much speculation, interest, and heated debate. Those who found it difficult to fathom keeping large constrictors as personal pets were quick to point their fingers at those that actually did. After all, history had already shown a loose correlation between the increasing popularity of reptile ownership and the quickening pace with which new reptiles had begun to appear around homes and neighborhoods in south Florida. And for the most part, the same species popping up unexpectedly in the wild were largely the same scaly faces that could be found behind glass at the corner pet store. For many, it was logical to deduce that the pythons now proliferating in the

Everglades were the latest example of a familiar cycle: impulse buys leading ultimately to accidental or intentional release.

In their defense, however, reptile enthusiasts and hobbyists adamantly refused to shoulder blame as a whole. Instead, many recognized that a small subset of irresponsible and inexperienced keepers could be found among them. If the pythons proliferating in the Everglades truly originated from the pet trade, it was not the fault of the overwhelming majority of diligent owners, but rather resulted from the negligent few who failed to provide proper enclosures, found themselves unprepared for the demands of ownership, or were simply too uninformed to know any better. Rather than malign an entire community of responsible reptile enthusiasts, they argued, energies would be better focused on better educating and regulating the irresponsible few.

But the irresponsible few remained difficult to find. Because people never believe themselves to be negligent keepers, no one ever stepped forward to volunteer their ineptitude or subject themselves to more stringent oversight. Nor did anyone ever plead mea culpa to the irresponsible handling of exotic wildlife. But even those who might have recognized some personal failure could still publicly divest themselves of any involvement in the growing python problem. Instead, they could clear their conscience by appealing to another popular scapegoat—the winds of Hurricane Andrew.

A Category 5 cyclone, Andrew dealt an infamously harsh blow to south Florida, culminating in the costliest natural disaster of its time. With peak winds of 165 miles an hour, the storm brutally punished communities in southern Miami-Dade County—flattening entire housing developments, snapping telephone poles and power lines, obliterating Homestead Air Force Base, and causing damage to Turkey Point Nuclear Power Plant. Andrew's winds also hurled a seventeen-foot surge of water inland, leaving roads, yards, and trees littered with an incongruent placement of boats, traps, floats, and other flotsam. In its wake, the storm left behind over one billion dollars of losses in agriculture, and a totality of devastation that was difficult for victims to fully grapple with.

It is known that Hurricane Andrew was also responsible for the release of many captive animals from zoos, research facilities, and private collections in south Florida. Reports abound regarding the storm's role in the appearance of rhesus monkeys, sacred ibis, Asian swamp hens, lionfish, and other species not formerly known to persist in the wild. One of the innumerable casualties reported in Andrew's aftermath was a nascent reptile wholesaling business nestled on the outskirts of the Everglades in the then-rural town of Homestead. Housed in a flimsy former agricultural grow house, the facility was flattened during the storm, reportedly sending its thousands of cold-blooded occupants—including hundreds of baby Burmese pythons—into the howling atmosphere. Launched deep into the Everglades, it is believed by many that some of the snakes not only managed to survive the ordeal, but would go on to become the true progenitors of today's problem population.

The "dispersal-by-tropical-cyclone" theory has been espoused by some and disputed by others. Contrary to the commonly accepted belief that today's feral population is the result of numerous intentional or accidental releases over the years, genetic studies of populations recovered from the wilds of south Florida have revealed a close kinship among the animals. Those wanting to lay blame upon the furious winds of Andrew suggest these findings, coupled with known patterns of past importation, are in keeping with a single, large-scale release. Critics of this hypothesis, however, note that genetic similarity might only suggest a similarity of stock imported over time for trade. Furthermore, the remote outpost of Flamingo, from whence the population appears to have radiated, lies impossibly far from the former site of the reptile breeding facility. Had the storm resulted in a catastrophic release of individuals, how did they only find themselves in a remote expanse of tangled mangrove swamp over 40 miles away?

The debate over how Burmese pythons were introduced into the Everglades has led some observers unnecessarily into the weeds. In truth, documented encounters with large constrictors on the loose in Everglades National Park and elsewhere—including Burmese pythons—long predate the arrival of Hurricane Andrew. Irresponsible keepers had most certainly been implicated in the escape of snakes in the past. And yet, it was also entirely plausible that hurricanes and tropical storms could aid the release of far more. Numerous foreign species were documented to be running wild around south Florida in the still aftermath of Andrew. Reasonable minds could easily entertain the notion that the current problem unfolding in the Everglades could equally well be the result of accidental releases over time, a consequence of natural disaster, or some combination thereof. And to many, the argument was moot. Regardless of whether they were intentional or accidental, releases were occurring. And either by the hands of middlemen or end consumers, large constrictors bought and sold as personal pets were somehow making their way into one of North America's most threatened habitats. The finer details regarding how they were introduced would never be known with certainty, but in truth, they weren't all that important.

The convoluted arguments, allegations, and theories regarding how these reptiles had taken a foothold in the River of Grass provided a convenient distraction from considering broader, more important, questions. Finger pointing provided an effective schoolyard tactic for evading basic questions about why pythons are imported in the first place. Attempts to assign blame or critique efforts to eradicate the current population provided an ample rug beneath which to sweep larger, more complex issues of regulation, enforcement, and ethics. And for some, aggressive posturing provided an effective diversion from yet another important consideration—how efficiently the snakes were ravaging south Florida's native inhabitants.

3
Breaking the Chain

Clay DeGayner and Joanne Potts spend a good deal of time traipsing through the dense tangle of trees found on northern Key Largo. As part of their research, the pair routinely takes morning treks into the canopied forests of the Crocodile Lakes National Wildlife Refuge and the Dagney Johnson Botanical State Park. On a recent visit to the area, I asked the now-retired DeGayner why he spends his time voluntarily dodging and weaving through thickets of poisonwood, gumbo limbo, pigeon plum, and mosquitoes. His curt reply was accompanied by a broad grin as he said, "It's fun."

Collectively, the refuge and park protect roughly 10,000 acres of land and water—much of it tropical hardwood forest. At one time, these "hammocks" were abundant along the Florida Keys, growing along a spine of higher elevation bisecting much of the island chain. Over time, however, they have been largely cleared—stripped bare to make way for the countless roads, homes, and hotels that now beckon travelers from afar. But the remnant patch of forest that persists on northern Key Largo is a last stand of sorts for the many plants and animals that still rely upon this ever-dwindling habitat.

The relatively conspicuous American crocodile, for which Crocodile Lakes National Wildlife Refuge was created, is afforded federal protection thanks to its limited numbers and range. Though fearsome in both appearance and size, it is easily recognizable, familiar to most locals and visitors, and is generally admired when appreciated from a distance. But the hammocks of Key Largo also conceal a suite of lesser-known species that, though far more imperiled, suffer a chronic inability to spark passion in the masses. As it turns out, most people are entirely underwhelmed by the rather indistinctive Schaus swallowtail butterfly. Few people fawn over the life history of the sluggish Stock Island tree snail. And virtually no one swoons over a chance encounter with the Key Largo cotton mouse. Imagine, then, the public perception of an endangered rat.

As its name implies, the Key Largo woodrat is endemic to the largest island of the Florida Keys and is found nowhere else on Earth. The bug-eyed, large-eared rodents can grow to nearly a foot in length and weigh over half a pound. Though clearly a rat in appearance, they *act* more like a storybook pig—building homes in the hammock understory of sticks and twigs scavenged from the forest floor. These distinctive structures are painstakingly built and maintained over time as a refuge in which to rest, feed, and nest. Unlike their more familiar, freeloading cousins, woodrats avoid human dwellings altogether, preferring instead to roam freely over every layer of the warm, humid hammock.

Within their stick homes the Key Largo woodrats again prove themselves distinctive from other rodents: females will typically only produce two fuzzy pups per litter, and average only two litters per year. This low reproductive rate, coupled with a short lifespan and the significant loss of viable habitat, has pushed the cinnamon-colored woodrat to near extinction. Some studies have estimated that, at times, fewer than 200 individuals can comprise their entire global population.

As perhaps the most critically endangered mammal in the United States, the Key Largo woodrat garners significant interest from the research community. Joanne Potts, a PhD graduate from the University of St. Andrews in Scotland, orchestrated a multi-year population sampling effort on Key Largo. But able to spend only a limited amount of time in the States, she relied heavily on the work of a handful of island residents. Chief among her team were Ralph and Clay DeGayner, brothers who long ago realized the dream of wintering every six months in the favorable trade winds of the Florida Keys. Yet unlike the retirement enjoyed by so many of their neighbors, Ralph and Clay volunteer a great deal of time, sweat, and money in service to the Key Largo woodrat. Their work with the reclusive species continued in April of 2007, when Potts visited the island to follow the movements of several woodrats that had recently been fitted with minute tracking collars. Armed with only an antenna and receiver in hand, she and Clay hit the hammock early in the morning of Friday the 13th.

The pair worked their way carefully through the familiar system of trails that narrowly finger through the forest, occasionally encountering large piles of limestone rock rising conspicuously from the hammock floor. Stopping, they trained their hand-held receiver on the rhythmic rise and fall of beeps emanating from a radio collar in close proximity. The strength of the signal betrayed the sanctity of refuge and revealed the whereabouts of the targeted woodrat. Taking a quick peek inside a rectangular, silver-sided trap nearby, DeGayner made an uplifting discovery: the female woodrat they were tracking was also carrying an unexpected surprise—a single pup dangling beneath. "Now we've seen everything," the pair quipped in disbelief.

The morning had turned out even better than expected. They had easily found two of their three collared specimens and had even had the fortune of finding an additional newborn. After a try at locating their final woodrat, they could return to the refuge headquarters to happily report the most recent addition to the population. The last individual of the morning would require a short drive. He had dispersed about 1.2 miles south—which earned him the distinction of being the southernmost woodrat on record at the time.

Having parked their vehicle nearby, DeGayner followed Potts as she punched through a dense tangle of trees, shrubs, and vines. Though the pair was well accustomed to the demands of work off-trail, this brush was particularly thick. Potts advanced slowly, following the audible guidance from her tracking equipment. Surprisingly, their woodrat was on the move—changing direction as they drew near. Odd behavior, it seemed, for a creature of the night.

Their hunt led them to an opening beneath the canopy overhead where sunlight filtered to the forest floor and better illuminated the limestone-studded landscape around them. In trying to keep up with the woodrat's erratic movements, the normally sure-footed Potts lost her balance in the brush and crashed clumsily to the forest floor. From this newly horizontal vantage point, she found herself staring directly at the undulating flesh of a very large snake on the move. Though right before her eyes, the encounter could scarcely be called "face-to-face"—neither the snake's head nor tail was immediately visible to Potts who, given the animal's girth, was rather disinclined to search for either.

Potts is a native of Australia, a country well known for hosting many of the world's most deadly animals. Over the years, perhaps, this had served to ingrain her with the healthy fear of snakes that reared its head that morning. The unexpected encounter with an unfamiliar serpent in the brush sent the usually level-headed academic into a self-described panic attack. With eyes wide open and through shortness of breath, she snappily told DeGayner of the snake and suggested leaving—her language peppered with expletives that left little question of her distress. "What a way for an Australian to die," she thought to herself, "from a bloody snake bite half a world away!"

Not surprisingly, Potts was not sold on DeGayner's next recommendation. "We've got to catch it," he insisted. The erratic daytime movements of their third collared woodrat now seemed to make sense—it was likely being digested somewhere within the active serpent, and removal of the snake was the only way to safeguard other woodrats from a similar fate. Not having much experience grappling oversized snakes, however, he immediately placed calls to his brother Ralph and two other researchers nearby. Ralph remembers receiving that call and hearing Potts in the background insisting the snake was some thirty feet long. He loaded the truck with a rake, shovel, and whatever else he could grab in a hurry and headed off to meet the pair in the hammock.

By the time he reached the scene a few minutes later, however, the reptile was already subdued. Reinforcements had arrived and helped snare the snake, which turned out to be the first wild Burmese python discovered on Key Largo. The responders took turns documenting the encounter with a series of photographs—one even including the smiling (though still white-knuckled) Potts (Figure 7).

Several days later, a necropsy of the snake would confirm the presence of a partly-digested woodrat still bearing a small, plastic radio collar. As it turns out, Potts and DeGayner had been unknowingly tracking the snake for several days. Had the python not ingested the tiny transmitter, it likely would have never been detected, leading others to wonder if more might be lurking in the island's hammocks.

The post-mortem analysis also helped confirm the final length of the troublesome invader as only seven and a half feet. Though certainly larger than any native snake, it was a far cry from the double-digits imagined by Potts. Still, one could speculate that Potts, having invested heart and soul in the well-being of the woodrats on Key Largo, actually experienced a subconscious premonition of something far more frightening than length. After all, gut content analysis would reveal one more unexpected surprise—the body of a second woodrat was found to have been recently consumed. For a species teetering on the brink of extinction, the threat posed by just one hungry python could be downright horrifying.

Across the planet, established species of nonnative microbes, plants, birds, fish, mammals, reptiles, amphibians, mollusks, and insects number in the thousands. In the United States alone, millions of acres are stifled beneath sprawling blankets of kudzu (*Pueraria spp.*), air potato (*Dioscorea bulbifera*), cogon grass (*Imperata cylindrica*), tamarisk (*Tamarix*), English ivy (*Hedera helix*), and Russian olive *(Elaeagnus angustifolia)*. What native lands escape being ravaged by photosynthetic competition must nonetheless contend with an endless parade of foreign ash borers, fruit flies, gypsy moths, long-horned beetles, ants, and other arthropods that defy easy categorization. Throughout the nation, streams, ponds, canals, and lakes are choked by the rapid growth and spread of hundreds of nuisance freshwater aquatic species—some of the more infamous being water hyacinth *(Eichnornia crassipes)*, purple loosestrife *(Lythrum salicaria)*, hydrilla *(Hydrilla verticillata*), Brazilian waterweed (*Egeria densa)*, and Eurasian water-milfoil *(Myriophyllum spicatum)*. The biological trauma clearly visible from the surface often mirrors change taking place within the water column below, where introduced crayfish, trout, eels, snakeheads, mussels, bass, and carp threaten to crowd out iconic American species with every successive generation. And for decades the Great Lakes have fallen under siege from a barrage of biological invaders that have adroitly taken advantage of artificial waterways, hitchhiked on the hulls of transient vessels, and been ejected in ballast water from passing container ships, including sea lampreys *(Petromyzon marinus),* round gobies *(Neogobius melanostomus),* alewives *(Alosa pseudoharengus),* quagga mussels *(Dreissena rostriformis bugensis),* zebra mussels *(Dreissena polymorpha),* and several species of water fleas. Damages from invasive species, and their subsequent costs of control, are thought to total a staggering $137 billion every year.

Even the farthest reaches of the United States are not immune to invasion. Avian malaria, transmitted via the introduced southern house mosquito *(Culex quinquefasciatus),* has facilitated the extinction of at least ten birds native to the Hawaiian Islands. Roof rats *(Rattus rattus)* and Norway rats *(Rattus norvegicus)* have been introduced to the Aleutian Islands of Alaska, where they prey heavily on nesting seabirds and seem well-poised to spread farther north. Free-ranging populations of introduced rhesus macaques *(Macaca mulatta)* and patas monkeys *(Erythrocebus patas)* now number in the hundreds in southwestern Puerto Rico, where they cause nearly $1.5 million in damage to fruit and vegetable crops annually.

Beyond our borders, the problem of invasive species is far more extensive. Lionfish (*Pterois volitans)* have spread throughout Caribbean waters like a hungry plague of locusts, stripping coral reefs of juvenile fish. Massive Nile perch *(Lates niloticus)* introduced into Lake Victoria and other African waterways have been blamed for the extinction or near-extinction of numerous endemic species of fish. Cane toads *(Bufo marinus)* introduced long ago in Queensland, now advance yearly to new reaches in Australia, where their ravenous appetite threatens smaller native species and their poison-producing glands kill potential predators. Eastern gray squirrels *(Sciurus carolinensis)* have overrun England, crazy ants *(Anoplolepis gracilipes)* have formed supercolonies on Christmas Island, caulerpa seaweed *(Caulerpa taxifolia)* thickly carpets the bottom in the cold waters of the Mediterranean, and giant African snails *(Achatina fulica*) have established themselves as serious agricultural pests in the far-flung reaches of China, New Zealand, and the Republic of Trinidad and Tobago. And on some island ecosystems, a never-ending battle against invasive rats has prompted the intentional release of the Indian mongoose in hopes of control—only to suffer an ecological bait-and-switch as our intended saviors become agents of destruction themselves.

Despite the seemingly endless catalogue of problematic organisms introduced around the planet, the list of snakes that have become established beyond their natural range has historically been, by comparison, relatively short. But scientists are beginning to piece together a more diverse portfolio of instances where snakes have quietly begun to make themselves at home in foreign locales. Today, dozens of nonnative populations are now known to be present around the globe, and more are being discovered all the time.

Though common throughout mainland Mexico, no sightings of the boa constrictor *(Boa constrictor)* had been made on the island of Cozumel during nearly 200 years of intensive surveys by myriad naturalists and herpetologists. That would soon change when, in 1971, an animal handler working on the production of the Mexican film *El Jardin de Tia Isabel* released an undetermined number of individuals around Palancar Beach. Eyewitness accounts cite the original stock to be anywhere between two and thirty individuals, but this much can be certain: at a minimum, there was at least one male and one female.

In subsequent years, boas have become a commonplace occurrence on Cozumel. Various sizes are regularly found, and a series of transect studies yielded a rough estimate of density: for every 100 hours of sampling, researchers would uncover two snakes. Though this search-to-find ratio may not strike one as particularly significant, it is important to remember that the number likely constitutes only a conservative estimate given the cryptic habits of boas. And for perspective, it is also important to remember that nearly 200 years of searching previously revealed none.

The southern watersnake *(Nerodia fasciata)* is a native of the southeastern United States. In the wild, most watersnakes exhibit bad temperaments and are likely to bite viciously if captured, while spewing streams of foul-smelling musk and feces in self-defense. Yet despite such infamy, they are occasionally available in the pet trade for purchase. Several different watersnake species have become established in California, and it has been postulated these localized populations began with releases by sorely disenchanted owners.

Today, the southern watersnake continues to thrive in the waters of Sacramento and Los Angeles Counties.

The southern watersnake is also established around Cameron County, Texas. Unlike those in California, it is believed this population originated in 1933 when a hurricane destroyed some cages belonging to the "Snake King," a local animal dealer. It was also suspected that this individual intentionally released snakes in the area both before and after the storm to create a steady supply of feeders for his captive cobras.

A small self-sustaining population of Aesculapian snakes *(Elaphe longissima)* has been established in North Wales in the United Kingdom for nearly forty years. Wolf snakes *(Lycodon aulicus)* from Southeast Asia now colonize islands in the Mediterranean Sea and Australia. A relatively new population of boa constrictors was reported on Aruba in 1999 and, after only a decade, now ranges across the entire island. And a colony of North American corn snakes *(Pantherophis guttatus)* has become established on the islands of Grand Cayman. And on Gran Canaria, one of the largest of the Spanish Canary Islands, authorities have captured and euthanized nearly 700 feral California kingsnakes (*Lampropeltis getula californiae*) since the late 1990s.

———

The discovery of Burmese pythons in the Everglades has made residents of south Florida only recently aware of the arrival of nonnative snakes. Most are genuinely surprised to learn that pythons are far from the first to reach the area. In truth, introduced serpents have been present and breeding in south Florida for more than three decades.

The Brahminy blind snake, reputed to be the most widely distributed serpent on the planet, was observed in the area of Miami for the first time in 1979. Reaching only six inches in length, the creature's dispersal has been aided both by its diminutive size and reproductive prowess. The snake is parthenogenic—part of an all-female species capable of reproducing without male fertilization.

Preferring a subterranean existence, the Brahminy blind snake is seldom seen on the ground's surface and is most commonly encountered when turning up soil. Burrowing beneath the ground, the snake feeds readily on the eggs, larvae, and the pupae of ants. It is this penchant for burial that leads many to surmise it has been introduced worldwide largely through the plant trade—earning the Lilliputian reptile its other common name—the flowerpot snake.

Owing to its secretive nature and small size, the Brahminy blind snake has proven to be one of our stealthiest invaders. Even today, most south Floridians are largely unaware of their presence and, even if encountered, will write them off as a harmless earthworm thanks to their awkward flailing and segmented appearance. Yet larger, more conspicuous snakes have also been established in south Florida for some time.

The boa constrictor, as we have seen, has become commonplace on the islands of Cozumel and Aruba. And since 1989, these large serpents have also become increasingly common in a popular park in the middle of Miami—the Deering Estate. Considered by many to be the crown jewel of the Miami-Dade County park system, the Deering Estate is huge by municipal park standards. The former residence of a wealthy pioneer, the estate sprawls over 450 acres of lush tropical forest and coastal habitat and affords visitors a picturesque view over the warm, sparkling waters of Biscayne Bay. Many locals treasure the park, set amidst a heavily developed residential area, as a remnant slice of natural Florida. But for years, unbeknownst to most, the park has also harbored a breeding population of boa constrictors. Between 1989 and 2005, ground crews captured nearly one hundred of them from the wilds of the park. The recovery of various sizes—from mature adults to juveniles less than one year old—led to the first report of an established population in the area in 1992.

Boa constrictors continue to be recovered into the present at the Deering Estate. Though the population appears to thrive, a dense complex of streets and subdivisions virtually isolates the snakes from the remnant freshwater Everglades and has likely helped to prevent the dispersal of boas beyond the park. It is perhaps the limited reach of this invasion that failed to draw any measurable interest from the local community. Despite nearly twenty years' notice, and the benefit of data collected both domestically and abroad regarding the invasive potential of the boa constrictor, the issue of nonnative snakes garnered little attention before Burmese pythons made their media debut almost a decade later. Consider as evidence that, even today, the private ownership of red-tailed boas remains largely unregulated.

With Brahminy blind snakes, boa constrictors, and Burmese pythons all established in the wild, south Florida is ground zero for nonnative serpents—boasting more species than anywhere else on the planet. But the paucity of similar invasions elsewhere leaves resource managers in south Florida to largely fend for themselves, having little precedent from which to borrow strategies in controlling the growing problem. And where snakes *have* established themselves in foreign locales, there is often little effort exerted toward control at all. Though few concrete examples exist on how to control invasive snakes, one important case study tragically illustrates the urgent need to try.

The brown treesnake (*Boiga irregularis*) was likely introduced to the island of Guam during the transfer of military equipment from Papua New Guinea to the U. S. territory following WWII. The native range of the species includes the coastal regions of eastern and northern Australia, extending up through the intricate mix of sea and land stretching westward from the Solomon Islands and falling just short of reaching the southernmost Philippines. They are slender in build yet boast bulbous heads sporting large, bulging eyes, appearing as if the hands of their creator may have squeezed excessively hard in rolling them to their three-foot lengths. Their catlike eyes, sinewy bodies, and prehensile tails provide definite advantages as they creep slowly through tree canopies in the dark of night.

The first brown treesnakes were encountered on Guam in the 1950s near the naval port. Over ensuing decades, the serpents began to colonize every available corner of the island in densities that sometimes rivaled those of all snakes found in the Amazon basin. Being dietary generalists, the snakes fed readily—devouring the diverse collection of birds, mammals, and reptiles that had evolved in the absence of similar predators on the island. "Island tameness" is the term used by some to denote the relative innocence that betrays such island residents. As vertebrate prey became scarce, an insatiable appetite led the snakes to sample rather unconventional fare. "We have records," noted several field researchers, "of [brown treesnakes] eating or attempting to swallow dog food, chicken bones, raw hamburger, maggot-infested rabbits, paper towels, spareribs, rotting lizards, ornamental betel nuts, larger conspecifics, human babies, dog placentas, and soiled feminine hygiene products."

The deleterious effect the introduction of this ravenous snake has had on both natural and human environments on Guam has been extensively documented. Brown treesnakes have been implicated in the precipitous decline of numerous island species, including the complete extirpation of eight native forest birds. Since the introduction, Micronesian kingfishers, Mariana fruit doves, Guam flycatchers, bridled white-eyes, Guam rails, rufous fantails, white-throated ground doves, and Micronesian honeyeaters have been completely silenced in Guam's inland forests. And because the brown treesnake is not a discerning predator, it is also the likely culprit in the recent loss of other forest organisms like the spotted-belly gecko, Mariana skink, snake-eyed skink, and two species of bats. Other research demonstrates the effects of the nonnative serpents extending well beyond the island interior and impacting a wider range of life along the shore, with disastrous consequences to white-tailed tropicbirds, brown noddies, and brown boobies.

Guam's ecosystems, now compromised, present the island with an uncertain future. Traditional food webs and niches have shifted, providing new opportunities for fellow invaders like shrews and geckos to disperse and multiply—in turn providing the remnant population of brown treesnakes with a continued stream of prey. The extirpation of numerous insectivorous birds and mammals has led to a population boom of spiders on the island and may also provide an easy environment for foreign insect invasions. Predation by brown treesnakes has also resulted in the widespread loss of traditional pollinators and dispersers on Guam, possibly altering the density and/or distribution of many trees and plants of importance on the island. Over a span of only forty years, the island of Guam has suffered an ecological collapse unparalleled in modern history. The brown treesnake, widely regarded as the single most important catalyst in this crisis, has also proven itself to be a noxious pest to human populations on the island.

Brown treesnakes now commonly inhabit Guam's urban areas, bringing them in close proximity to residents and visitors. Brown treesnakes are mildly venomous, with an aggressive predisposition when threatened. Though envenomation is not generally life-threatening and produces only moderate pain in healthy adults, very young victims can require hospitalization and intubation. The rapacious appetite of the brown treesnake, not being limited to the forests of Guam, often lures the invaders to feed upon young domestic stock and fowl, much to the chagrin of owners. Being adroit climbers, they are also known to cause frequent power outages on the island as they maneuver between utility poles and transmission lines. It has been noted that this phenomenon does not help the fledgling tourism industry of Guam, nor does the prospect of visitors encountering large, predatory serpents during their holiday.

The grim reality of life amidst a thriving population of brown treesnakes is not lost upon island communities elsewhere. The ease with which these animals exploit traditional transportation pathways—particularly aircraft—has been proven repeatedly. Multiple sightings reported from the islands of Saipan and Oahu have communities throughout the Pacific on a constant state of high alert. Trained rapid response teams stand ready to mobilize in the event snake sightings are reported. Eager to avoid similar impacts to their environment, industry, quality of life, and personal safety, governments now undertake exhaustive measures to keep snakes from entering their jurisdictions—often at great cost and effort. Expensive barriers, constructed of either concrete or cloth, are erected and maintained around ports and natural areas in a hopeful bid to physically exclude the snakes from limited areas. Nighttime perimeter surveys are conducted by researchers and security personnel in the hopes of sighting occasional nocturnal transgressions. Jack Russell terriers are employed by the United States Department of Agriculture as detector dogs in an effort to ferret out stowaways from outbound and incoming cargo. It has been estimated that the combined cost of interdiction efforts, endangered species recovery programs, long-term research, and losses to businesses and agriculture from the brown treesnake hovers around $5.6 million annually.

Over the course of only a few decades, the brown treesnake has dramatically altered life on the island of Guam and mired communities throughout the Pacific in a never-ending game of wait-and-see. In summarizing the chaos, one team of researchers writes, "It is difficult to identify another introduced species anywhere on earth that has had such a comprehensive impact." But long before the mushroom cloud of extinction became visible on the landscape, the chaos was likely to have started humbly with only a handful of wayward snakes. And, as it turns out, the path to irrevocable extinction begins with the loss of only one or two individuals. Like, for example, the pair of woodrats found digesting in the innards of one Burmese python.

With the suspicion that pythons had taken a foothold in the wilds of south Florida came much speculation regarding their potential impact on the natural ecosystems of the Everglades. Based solely on their size and reputation as skilled predators, many believed their proliferation could disrupt the ecology of the area in a manner previously unimagined. Direct predation upon native animals, particularly a subset of the 22 federally-listed threatened and endangered species found in the national parks of south Florida, became one of the foremost concerns. For over a century, the establishment of hundreds of nonnative plant and animal species had already subjected south Florida to a slow torture by a thousand cuts—priming it, perhaps, for the arrival of that one catastrophic species capable of causing complete collapse. Some questioned whether the Everglades had become the newest stage for the unfolding of a drama reminiscent of Guam's plight.

Perhaps the only thing less pleasant than encountering a 13-foot python in the wild is gutting one post-mortem in a laboratory. Yet that is the unenviable task of Skip Snow, a park biologist with Everglades National Park who has worked on the Burmese python issue from its earliest chapters. Snow now estimates that he has personally performed necropsies on roughly half of the 1,700 pythons removed thus far from in and around the park. His methodical examination of individual specimens in his laboratory has provided some of the most important data yet in understanding the impacts of their presence in south Florida. He has earned a tremendous amount of fame and respect both for his encyclopedic knowledge of the facts and his thoughtful commentary on the many facets of the issue.

In the laboratory and on the job, Snow is the consummate scientist. He is deliberate and calculated in his attempts to ask the right questions. He is careful to use the best available practices to obtain the answers he seeks. And he is consistently reluctant to draw premature conclusions from the data he subsequently acquires. When pressed to describe the crux of his work with Burmese pythons over the past decade, Snow is quick to reply, "We just told people what was going on."

Like so many career professionals in the agency, Snow enjoys fond childhood memories of traveling to various national parks with his family. He traces his desire to work for the National Park Service, however, to a little-known monument outside Flagstaff, Arizona. He remembers how, during a family visit to Walnut Canyon, he joined a ranger-led walk that circled the rim of the steep-walled gorge below. After expounding upon the arid terrain of the area and the historic dwellings of past inhabitants found alongside the trail, the park guide concluded his presentation directly across the deep chasm from the now-distant visitor center. Needing to return to his office, the ranger bid goodbye to the group and, rather than continue around the rim, opted instead to take the most direct route possible. He hopped the nearby railing and quickly disappeared into the seemingly bottomless canyon below. Snow still remembers watching as his distinctive flat hat descended out of view step by step. "What better job is there than *that*?" he thought.

From that moment on, Snow decided he would demand two things of his future career: he wanted to work outside, and he wanted to somehow work with animals. He would go on to study zoology at Miami University in Oxford, Ohio, where he proceeded to also earn a graduate degree in environmental sciences with a concentration in park planning and management. Following his studies, Snow embarked on a career with the National Park Service that began in Denver, Colorado, but soon transitioned to new positions at Mount Rainier and Theodore Roosevelt National Parks.

In 1988, Snow traveled to the humid, subtropical wetlands of south Florida to accept a new job as the backcountry management specialist for Everglades National Park. The promise of paddling the twisted waterways of the mangrove coast as part of his official duties was a big draw, as was the opportunity to gain experience in the relatively new discipline of fire management. Everglades was the first national park in the nation to implement the use of prescribed burns as a necessary means of perpetuating vital ecological processes. Over subsequent years, the charring of vast acreages of Everglades pinelands would gradually help transform the entrenched public perception of wildfire as a threatening foe to that of being a largely misunderstood benefactor.

During his tenure with the park, Snow's original job description slowly morphed into that of an Everglades polymath. During the last two decades he worked in virtually every environment of the ecosystem, from the monitoring of sea turtles and other marine resources to the reintroduction of birds formerly extirpated from the park's uplands. Wearing many hats, Snow conducted regular manatee surveys, performed administrative project reviews, participated in resources management planning, and helped oversee activities in the park's vast wilderness area. As an ancillary duty, he began working a variety of fire assignments—first smothering flames on the front lines as part of a hand crew before eventually working as a burn boss and fire behavior specialist trainee during prescribed ignitions. Snow became simultaneously a field biologist, contract agent, public outreach specialist, and administrator.

In 2003, shortly following the much-publicized alligator/python duel at the Anhinga Trail, Snow began gathering as much data as possible about the Burmese pythons being increasingly encountered in the area. He began compiling and mapping every known report and recovery across the south Florida landscape. He conducted regular, nighttime surveys along park roads and nearby canals. And he began to amass an impressive collection of post-mortem data from pythons dissected in his laboratory. At long last, he found himself out in the field frequently, interacting with impressive wildlife, and making inroads into subject matter that, though familiar in language and concept, was itself relatively unknown. In a manner beyond his wildest imagination, Snow was fulfilling his childhood aspirations in the wilds of the Everglades.

Early speculation held that necropsies on captured pythons, like those performed by Snow, might help discriminate between newly released snakes and truly feral individuals. In theory, recently released captives would be more likely to have a healthier appearance and retain high levels of fat, since snakes are not likely to suffer injury in captivity and are typically well-fed and restricted in their movements. Conversely, wild individuals that spend time roaming the landscape and are subjected to irregular feeding patterns might appear far leaner and perhaps bear outward signs of physical trauma from a life on the lam. Thus it was hypothesized that examining the overall condition of pythons recovered in the Everglades could help explain their relative place of origin.

In practice, however, most of the snakes captured showed little evidence of trauma and generally appeared to be in good condition. And differences in fat levels were not immediately conspicuous. It seemed as though all the pythons, regardless of whether they originated from the freshwater marshes of the park or a townhouse in Miami, were thriving beautifully in the wild. Snow suggests that in the future, the dataset being compiled may be useful in demonstrating an overall improvement or decline in the body condition of the python population over time, perhaps resulting from changes in the available prey in the area, or a shift in environmental conditions.

Necropsies have, however, helped inform our understanding of the life history of pythons in south Florida. They have shed light, for example, on the reproductive prowess of the species in the Everglades. Though the recovery of various sizes provided initial suspicion that the pythons were breeding in the park, the laboratory examination of recovered females provided repeated, sobering evidence of reproduction—body cavities packed to the walls with fertile eggs. Gravid females have been recorded bearing clutches ranging from 21 to 85 eggs, suggesting their fecundity in south Florida is on par with what has been documented in their native range. And by demonstrating that egg-bearing females are most likely to be encountered in the Everglades between March and May, necropsies have also helped better illuminate the timing of reproductive cycles in the New World.

Perhaps the most valuable information to be presently gleaned from post-mortem examination comes from "gut content analysis," the examination of digestive cavities to identify recent meals. Between 2003 and 2006, Snow and a team of researchers examined the stomach and intestinal contents of 56 Burmese pythons captured in and around Everglades National Park. While their goal was to document the individual species targeted as prey, the partially digested remains of some could only be identified to the genus, order, or family thanks to an advanced state of decay.

A total of 54 animals were exhumed from the innards of the study subjects. Of these, roughly 70% were mammals: rodents, raccoons, squirrels, cats, one opossum, and a juvenile bobcat. The balance of prey was almost exclusively comprised of birds: a limpkin, a white ibis, an American coot, pie-billed grebes, and an assortment of winged creatures whose specific identity could not be determined (Figure 8). It appeared the dietary preferences of Burmese pythons in the Everglades were mimicking those in their native range, and that warm-blooded animals were most at risk of predation.

In the months and years that followed the study, an unending procession of necropsies expanded considerably the list of potential prey. Pythons were found to be digesting an increasingly diversified portfolio of birds—25 species in all have been recorded thus far. The abundance of wading birds exhumed—great egrets, red-winged blackbirds, little blue herons, anhingas, house wrens, and snowy egrets—reveals a predilection for hunting at the water's edge. And rails in particular were shown to constitute a significant part of the Burmese pythons' diet, a troubling fact considering the manner in which close relatives had been decimated by introduced predators on Guam.

Domestic chickens and ducks from nearby farms and residences were also identified in the innards of captured pythons. The consumption of a great blue heron suggested that even large birds were at risk, and one python recovered from deep in the freshwater marsh was later found to contain the puzzling remains of a magnificent frigate bird—a coastal-dwelling species with a considerable seven-foot wingspan. Predation on rabbits was documented, and the discovery of undigested hooves added white-tailed deer to the serpentine menu. The consumption of alligators, so novel a concept in 2005, was now being documented repeatedly—the largest of which was roughly six feet in length. And the discovery of two federally endangered species, the Key Largo woodrat and the American woodstork, brought renewed gravity to the potential impacts of the invasive snakes.

Evidence suggests that over the course of a lifetime, Burmese pythons are capable of consuming most warm-blooded terrestrial vertebrates of the Everglades. This becomes particularly illuminating when coupled with other digestive realities. Generally reptiles are far more efficient at assimilating prey than most warm-blooded organisms. Encumbered with the need to maintain a constant body temperature, mammals and birds typically lose the vast majority of their caloric intake (as much as 90% in some cases) to mere heat. It is known, however, that captive Burmese pythons can utilize approximately 30–40% of their prey to support growth and reproduction. Though the energy efficiencies of free-ranging pythons are not fully understood, it can be safely assumed that greater opportunities for activity and movement would leave less energy for development. Drawing upon laboratory data and field measurements, researchers constructing a hypothetical energy budget found that wild pythons might only utilize as little as 4% or their caloric intake for growth.

Put another way, building and maintaining just one pound of python in the wild could require approximately 25 pounds of food. Capitalizing on both this inferred rate of efficiency and the present understanding of dietary preference, one could begin to construct a theoretical representation of the diversity and quantity of native species reasonably expected to be consumed during growth from birth to adulthood. In one particular iteration of this exercise, researchers postulated that the required calories necessary to drive the growth of a single 13-foot python could require the consumption of literally dozens of birds, scores of small mammals, and perhaps even several large crocodilians. Extrapolate that across just a single clutch of 85 baby pythons, and the numbers begin to dizzy the imagination.

Direct predation of native species is, by far, the best documented and understood impact resulting from the introduction of Burmese pythons in the Everglades. But given the known strategies, haunts, and abilities of the invader in its native range, scientists have speculated about the possibility of additional impacts that are far more difficult to both demonstrate and assess.

It is known, for example, that Burmese pythons will readily occupy holes, burrows, and subterranean refugia, and it has already been demonstrated that they continue this practice in the New World. Consequently, it has been suggested that this character trait may put the python in direct conflict with a number of native cavity-dwelling species in south Florida—particularly if they prove amenable to occupying the subterranean haunts of the gopher tortoise (*Gopherus polyphemus*). A state-protected species and an engineer of cavernous underground burrows, the gopher tortoise shares its refuge with a host of other symbiotic animals. One burrow can become an ecosystem unto itself, attracting dozens of other vertebrates—owls, mice, rats, skunks, wrens, raccoons, frogs, toads, turtles, opossums, foxes, rabbits, and quail. Snakes, often seeking either prey or warmth, will also frequent the burrows, including Eastern diamondback rattlesnakes and the federally-threatened Eastern indigo snake.

Though no Burmese pythons have yet been known to infiltrate these burrows, there is little to suggest they won't. And when they do, there is even less to suggest they wouldn't feast readily upon the warm-blooded residents within. And as for any reptilian competition it might encounter, it is unclear whether or not a hefty python would be as willing to share quarters with a smaller roommate. Evidence reveals this drama might already be unfolding:

staff at the Rookery Bay National Estuarine Research Reserve recently snapped photos of a shed left behind at the entrance to a gopher tortoise burrow by what appears to have been a departing Burmese python (Figure 9).

Noisy, swarming flocks of nesting birds have become an iconic symbol of the Everglades ecosystem, though over the past century their numbers have dwindled under pressures exerted by avaricious plume hunters and abrupt changes in water management. The large rookeries that still persist into the present might also have to deal with an additional complication posed by pythons. Given the ease with which the snakes can maneuver their ponderous bulk into the treetops, and having discovered their zeal for the taste of avian prey, it is only logical to surmise that pythons would find large groups of nesting birds gastronomically attractive. In a game of numbers, a large colony might easily absorb the cost to settle the hunger pangs of a single python. But nesting birds may react to repeated disturbance by several threatening pythons in much the same way as they do to hunters and unfavorable water levels—by abandoning their nests altogether. The yearly nesting success of wading birds is critical to the recovery of populations in south Florida, and a marauding band of Burmese pythons might not help.

While some continued to caution about the possible effects of pythons on south Florida's natural areas, others began questioning how their presence might hit closer to home. Florida is, of course, a huge travel destination, and tourism remains the state's biggest industry. Could the python have some measurable impact—either as a catalyst for the loss of charismatic species that provide a draw for visitors, or perhaps as a deterrent to those who might mentally envision pythons draped precariously atop every swaying palm tree on South Beach? Worse yet, might there be perfectly valid concerns about the safety of visitors—particularly children—resulting from the proliferation of large, ambush predators to which they may be neither accustomed nor aware? Some have attempted to downplay the effect, suggesting that the biggest threat to visitors stemming from pythons is likely as a road hazard. And indeed, the thought of running into a 12-foot python stretched across US 1 at 50 miles an hour is a scary proposition!

Some residents on the periphery of Everglades National Park had already had chance encounters with large snakes and, in some cases, were losing domestic animals in the process. One of the pythons outfitted with a radio transmitter was tracked to a residence just outside the park boundaries, where it was found to have recently consumed a neighbor's pet goose. Snow would later recount how, as a result, he used his personal vehicle to deliver a surprisingly diarrheic replacement to the distraught homeowner.

Still, since they had lived in the realm of panthers, alligators, and crocodiles for so long, there was little to suggest residents couldn't adjust to a life among pythons. And so far, south Florida was still a far cry from Guam. In fact, when viewed through the lens of ecological theory, the Everglades stands in stark contrast to a remote Pacific island. The same rules that can help facilitate environmental collapse in one area do not apply uniformly across the globe. Where isolated species might suffer "island tameness," the fauna of the Everglades does not. Unlike the lost species of Guam, which in some cases occupied very limited ranges, the Everglades ecosystem is far more expansive and most native animals—with notable exceptions, particularly along the Florida Keys—are far less specialized. And though the River of Grass is in many ways already quite diverse, there seems no shortage of available niches to be occupied. And good thing—unlike an island in the middle of the ocean, the area suffers a constant immigration and absorption of new species. Over the last century alone, the Everglades have assimilated hundreds of introduced plants and animals. So what could be wrong with adding just one more?

4
One of the Many

Crandon Park, situated on the northern end of the island of Key Biscayne, boasts one of the most popular beaches in Miami-Dade County. On any given weekend, you will find the full stretch of tan-colored sand awash in both surf and sunbathers. Trumpets and timbales blare from the many cabanas and shelters throughout the park, broadcasting an endless rotation of timeless salsa favorites. Kayakers vie with kite boarders for dominance over patches of shallow water while, on land, lifeguards keep watch from a network of aged, sun-bleached towers. Up near the dunes, throngs of barefooted revelers pursue the full gamut of traditional ocean-front games—volleyball, paddle ball, Frisbee, and sand sculpture.

Not too far from the salt-encrusted chaos of the shore lies a far less visited and largely unknown corner of the popular park. Crandon Gardens, as it is called today, is a large, fifty-acre expanse of lakes and lavish landscaping. The area once served as the home of the county zoo, founded in 1948 with a motley assortment of exotic wildlife. Over time, the facility grew to comprise some 1,200 animals, including rare white tigers and Indian rhinos. Following the devastating effects of Hurricane Betsy in 1965, the zoo was moved to an inland location, where it continues to operate today. The former site has subsequently been reinvented as Crandon Gardens, and the area is now open to tours by the visiting public. And though a zoo no longer operates on the grounds, those who spend time in the garden quickly learn there is still an abundance of exotic wildlife to encounter.

I first visited the garden in 1997. At the time, I was working for the local parks system and had been sent to meet with staff in our regional office nearby. The garden was in the midst of a major restoration and was not yet open to the public. I arrived quite early in the morning—well before anyone else had walked the area that day. As I walked through the gate in the tall, black, chain-link fence that girdled the area, I was overcome by succeeding waves of surprise and awe at what greeted my arrival.

With every step forward, the brilliantly verdant grass before me came alive. Formerly motionless, scores of hefty green iguanas began to stir nervously as I approached. Every step revealed a greater number than before, each growing increasingly apprehensive as I loomed larger before them. Drawing nearer, I noted the myriad sizes and hues in which the animals presented themselves. Then suddenly, as if I had crossed some threshold of safety which only they could see, the entire reptilian mass bolted upright and, in unison, began running on two legs toward a lake nearby. "Tiny velociraptors," I thought to myself, recounting the familiar Hollywood scene from *Jurassic Park*. Upon reaching the water, each iguana dove headfirst and disappeared beneath spreading, concentric circles rippling on the surface.

The scene would repeat itself several times that morning, as I startled new-found groups of iguanas moving from one area of the garden to the other. That day I would also encounter dozens of reclusive spiny-tailed iguanas, witness the frantic darting of resident "jungle-runners," and watch clusters of southern brown basilisks reenact biblical lore as they dashed quickly on the surface of the water without sinking below. Though the zoo was long gone, the marvels of a reptilian menagerie were still on daily display.

Over the last century, it appears an awareness of foreign species has gradually entered the consciousness of most Americans. Across the country, it is commonly recognized that the plants and animals of a given region can be segregated into two main groupings: those that are native to an area, versus those that are not. Furthermore, there seems a growing understanding that the arrival of alien species sometimes brings unwanted, and almost always unexpected, consequences. The very term "exotic species" is now firmly entrenched in our lexicon, and awareness of their impacts augments in step with the proliferation of the very organisms it describes. And our disdain for the arrival of new exotic species grows in lockstep with our continued experience with them.

Generally, those organisms that either have occurred or might occur in a given area as a result of purely natural processes are understood to be "native." In contrast, a nonnative species (alternatively known as an "exotic," "introduced," or "alien" species) is one that is introduced beyond its natural range either directly or indirectly through the deliberate (or accidental) actions of people. Of course, taken at face value, these definitions presume that human beings are themselves alien—having no place in the natural world as biological vectors, no authority bestowed by Mother Nature to act as agents for dispersing other organisms throughout the biosphere. It is well known, however, that our ancestors long ago carried crops of nutritional, medicinal, and religious importance across entire continents as they migrated over millennia. Were they, in fact, seeding the continent with harmful invasive species in their wake?

In truth, differentiating between native and nonnative species often requires a certain degree of interpretation. The influence of people in distributing harmful species grows in proportion with how significantly we surpass the natural constraints that limit the reach of other species on the planet. Specifically, it is our new-found ability to harness external sources of power—to move quickly across the continents, render wholesale changes upon the landscape, and purposely reposition nature to suit our many needs—that now places us squarely outside the ancient rhythms of biology. It is one thing to walk, as other animals might, across a beach dune with sandspurs hitching a ride on your sandals—yet it is quite another to leave the beach, board an airliner, and fling those same sandspurs across the ocean at 500 miles per hour.

Still, discerning between native and exotic organisms can be a somewhat messy endeavor, particularly in poorly studied areas or those with long histories of human traffic. Native species are typically identified in baseline inventories cobbled together from the disparate historical accounts of botanists, naturalists, collectors, explorers, and pioneers. Our understanding of native landscapes, therefore, is only as comprehensive and accurate as the information upon which it is based. And discerning the origins of suspected introduced species is often equally fraught with uncertainty—relying upon the acquisition and review of shipping logs, manifests, ledgers, permits, and other obscure, and sometimes spurious, resources. Indeed, organisms sometimes prove so difficult to categorize as either native or exotic that academics have recently coined a new term to reflect their obscure origins—"cryptogenic."

Fortunately, most introduced organisms at least appear to behave themselves. That is, they remain confined to their intended purpose, be it a subject of research, a colorful garden planting, a valuable staple crop, a roadside attraction, or as the resident of a private aquarium. Should any number escape either their confinement or cultivation, however, the possibility exists that they might become "established." That is, they might form a feral, self-sustaining, reproductive population. And where introduced species become established, there is the additional chance that the population might ultimately prove "invasive"—spreading beyond a localized area and into new habitats where they begin to interplay with native biota and cause some measure of harm. Invasion biologists have long adopted the "rule of tens" as a rough formula for this dynamic: of those species introduced into a new area, ten percent are likely to escape; of those, ten percent are likely to become established; and of those, ten percent are likely to become invasive. It is this latter category, the invasive exotics, that poses the greatest concern to scientists and resource managers, as the ultimate impact of these organisms can prove both largely unpredictable, and sometimes, devastating.

Despite the nuance of invasive species terminology, and the uncertainties inherent in discerning their presence, modern invasions are somewhat easier to identify across the terrestrial landscapes of North America. Not only have many of these ecosystems been the focus of intense study, but the relatively clear timing of European contact provides a convenient—albeit arbitrary—demarcation point for identifying foreign introductions. And the emerging field of paleoecology continues to provide new insights into the prehistory of natural systems.

Similarly, the wilds of south Florida have been scrutinized by human eyes for centuries, and continue to be subjected to the newest research methodologies. The area hosts a mélange of organisms—some presumed native, some cryptogenic, and others clearly exotic—that continues to grow, with every passing year, in both number and diversity. And yet, despite the nearly clockwork appearance of new, high-profile species of concern, the net effect on popular opinion seems almost imperceptible. Not unlike the immigrant experience shared by so many in south Florida, new arrivals—regardless of taxonomy or place of origin—are all quickly assimilated and subsequently afforded little concern.

While Crandon Park may boast a particularly impressive collection, all of south Florida has become a virtual Ellis Island for nonnative lizards. For decades, an ever-growing portfolio of legged reptiles has quietly infiltrated the landscape, exploiting every possible nook and cranny available in urban and natural environments. During this insidious invasion, many have remained cryptic and reclusive, using camouflage to deflect the attention of human eyes. Others have simply remained well-hidden within the confines of deep burrows, or within the porous structures of limestone rock. Others, growing bolder as years pass, now bask brazenly along canal banks or openly colonize residential yards. Over time, these exotic lizards have become so commonplace they are now regarded by many residents as iconic symbols of life in the Sunshine State.

Perhaps because they now saturate every available patch of real estate, the sight of a dozen lizards basking nearby doesn't raise the eyebrows of most locals. Rather, it is often through the eyes of unsuspecting visitors from abroad that one begins to truly appreciate the abundance with which lizards can be found. Unlike the temperate landscapes from whence many visitors hail, the subtropical climate of south Florida seems to breed cold-blooded masses that scurry both overhead and underfoot. Most visitors are likely to have their first close encounter with a brown anole (*Anolis sagrei*) basking on a heavily trafficked sidewalk. Where they occur, brown anoles are conspicuous, bold, and simply inescapable in their abundance.

The brown anole has proven to be a highly adaptable species, having spread throughout three-fourths of the peninsula since its arrival to mainland Florida sometime in the 1940s. It is, without a doubt, the most common and successful reptile to invade the area to date. Though originally the result of multiple introductions involving two separate subspecies, extensive interbreeding between the two has effectively diluted the distinguishing characteristics of Florida's present-day population. Today, this ground-dwelling, rust-colored anole thrives equally well in urban environments and disturbed natural areas. And every year, it appears, the population extends its reach slightly farther north.

Visitors and residents alike, though, tend to pay little attention to the details of any one species. Rather, any legged reptile that is clearly neither a crocodilian nor a turtle gets lumped into the broad category of "lizards." The more observant among them might even differentiate between the sleek anoles that patrol gardens and sidewalks during the day, and the stockier "geckos" that perform feats of daring acrobatics on ceilings and doorways in the evening. Yet more than merely noting quantity, it is the trained, attentive eyes of field biologists that have brought the extent of Florida's exotic reptile problem into greater relief.

In scientific practice, reptiles are scarcely studied alone. Instead, they are lumped together with amphibians under the discipline of herpetology, which literally means "the study of crawling things." The bias of taxonomists from long ago continues to intertwine the study of two very distantly related groups of life. This duality of practice has extended through the work performed by herpetologists in the wilds of south Florida over the past century.

In 1875, Edward Cope reported the presence of the first exotic amphibian discovered in south Florida—the diminutive greenhouse frog (*Eleutherodactylus planirostris*). Many years later, biologist Leonhard Stejneger would report the occurrence of two diminutive species of gecko as the first nonnative reptiles in south Florida in 1922. A subsequent survey over 40 years later reported 16 reptiles and amphibians known to be breeding in southeast Florida, with several additional species being of unknown status. An ecological assessment on the "herpetofauna" of south Florida conducted in 1983 by Wilson and Porras reported 20 nonnative species established and breeding in south Florida, including the first known snake invasion in the United States—the Lilliputian Brahminy blind snake. By 2004, Meshaka, Butterfield, and Hauge had documented 40 introduced reptiles and amphibians in the state and noted that the rate of successful introductions had clearly been increasing over time.

Every herpetological inventory provides insight into how newly introduced reptilian and amphibian life is spreading across south Florida. The cane toad (*Bufo marinus*)—a formidable, half-pound toad the size of a small dinner plate and armed with toxin-filled glands—had expanded readily throughout the southeastern coast of Florida. The Cuban treefrog (*Osteopilus septentrionalis*), appearing originally in Key West, subsequently marched up the island line and successfully colonized the mainland. And the greenhouse frog first discovered by Cope in 1875 eventually occupied the entire state, likely aided by a handy reproductive shortcut—the young hatch from eggs fully formed, bypassing entirely the usual tadpole stage of other frogs.

Still, successful invasion by foreign amphibians is relatively rare. For whatever reason, the establishment of nonnative reptiles is notably far more commonplace. And introduced lizards seem to fare particularly well. Brown basilisks (*Basiliscus vittatus*), dubbed "Jesus Christ lizards" for their ability to run across the surface of water when startled, have established colonies along canal banks throughout Miami-Dade and Broward counties. Populations of curly-tailed lizards (*Leiocephalus carinatus*), named for their flamboyant rear appendage, proliferate amidst the refuge afforded by the area's abundant limestone rock structures. Where their numbers flourish, the conspicuous croaking of foot-long tokay geckos (*Gekko gecko*) has become a cacophonous evening ritual. Eye-catching rainbow whiptails (*Cnemidophorus lemniscatus*)—with their spasmodic movements and beautifully iridescent scales—flourish incongruously near city basketball courts, perhaps aided by their ability to reproduce asexually. And a full suite of anoles of every size, each bearing a distinctly colorful throat fan, has occupied every imaginable niche in urban south Florida.

The periodic herpetological surveys undertaken by biologists do more, however, than merely document the movements of invading reptiles. Each also investigates the means by which species were introduced to the area. One such inventory identified three possible avenues for introduction: accidental importation, accidental escape from animal dealers and importers, or the intentional release of animals.

The globalization of commerce and the ease with which goods now circumnavigate our planet provide numerous pathways for the accidental introduction of nonnative species. Ships traversing the high seas bring with them unintended cargo—encrusting organisms on hulls and anchor chains, and all manner of free-swimming life in ballast water. The constant routing of ornamental plants, fruit trees, and produce between countries provides ample cover for the diminutive weevils, beetles, flies, and fungi that often hitch a ride undetected on plants that, in turn, may prove invasive at their destination. Larger organisms can sometimes escape the weary, unaware, or unconcerned eyes of importers and exporters. Under a lax watch, frogs, lizards, snakes, rats, and birds are afforded free passage to unsuspecting ports of call. And the introduction of the brown treesnake through the repositioning of military equipment suggests that even the act of war spawns pathways for new invasions.

Though accidental importation has clearly resulted in the introduction of numerous harmful species, the deliberate importation of wildlife has greatly exacerbated the problem through the egregious release of many more. One study of introduced reptiles and amphibians in south Florida revealed that approximately 56% of established species resulted from releases by animal dealers or purchasers. And while commerce has generally resulted in the introduction of more diminutive animal species with limited ecological impacts, the intentional trade in exotic wildlife has released considerably larger organisms capable of causing greater environmental harm. The plague of toxic cane toads that now terrorizes squeamish south Florida homeowners and devours native frog species is the direct result of escapees and intentional releases from several animal dealers in the area. Those who enjoy a chance encounter with the powerful jaws and menacing gaze of a Cuban knight anole (*Anolis equestris*) can thank the University of Miami student who, in 1952, first introduced the species on the grounds of the Coral Gables campus following a research project. And the Nile monitor (*Varanus niloticus*), Africa's longest lizard and a favorite import among collectors, now freely roams residential areas in Florida's west coast community of Cape Coral, where they can potentially grow to lengths of eight feet, and feed readily on snakes, frogs, rabbits, goldfish, ducks, other lizards and, presumably, anything else they can fit between their jaws.

The periodic surveys of nonindigenous species in south Florida often touch upon the potential effects stemming from their arrival. In aggregate, the introduced herpetofauna of the Sunshine State could yield a variety of impacts. As illustrated by the proliferation of exotic anole species, new arrivals could compete directly with native wildlife for food, water, shelter, and space—possibly displacing them outright. Worse yet, some ravenous imports could feed upon copious numbers of native organisms directly. Other impacts, though more subtle in nature, could prove equally damaging over time. Certain species, thanks to peculiarities in their biology, have proven distasteful to potential native predators, resulting in an energy "sink" whereby nutrients and energy consumed by the invader are never returned to nourish the food chain beyond. And in select species, hybridization (or even the attempt to hybridize) has the potential to disrupt the genetic integrity or stability of native populations.

Given the profusion of exotic lizards known to have taken a foothold in south Florida, naturalists rightly question why nonnative snake species have not also followed suit. Given their Houdini-like prowess, one might certainly expect serpents to make occasional, daring escapes from area pet shops, storage warehouses, breeding facilities, and private homes. And they have.

During the 80s and 90s, local newspapers ran frequent published pleas from private owners requesting assistance in finding the likes of Abby, Floydie, Sadie, Oglethorpe, Leviathan, and Jake the Snake—all escaped pet pythons enjoying a taste of personal freedom. Some made a daring jailbreak by forcefully contorting themselves through their cages. Others, enjoying free range within the home of their keepers, simply nudged screens off windows or slipped away through an open door nearby. One even crawled out the open window of a vehicle on Key Largo parked at the appropriately named Happy Vagabond Campground. Some enjoyed a sojourn of only one day; others traveled for weeks. One managed to enjoy a five-month-long walkabout before finally being repatriated. Some of them had escaped before and yet, when found, were returned to their owners yet again.

Not every python was found, of course, nor was every escape reported. Unexpected encounters with large, wayward snakes became increasingly common. Soon Burmese pythons, reticulated pythons, royal pythons, ball pythons, African rock pythons, and boa constrictors were being found loitering in mall parking lots, taking up residence in carports, nestled in pickup truck engines, emerging from toilets, perched in shade trees, and terrorizing condominium owners in the ritzy enclave of Key Biscayne. One 100-pound python was found slithering across a busy Hollywood street; another was discovered in a Lake Worth neighborhood after the 12-footer treed a noisy stray cat. They were being killed by hunters in camping areas and found by visitors to popular area attractions. One royal python was even found harassing a panicked caged parrot on someone's back porch.

The frequency of such encounters was troubling, and the colossal size of the snakes being found continued to grab headlines. At one South Miami construction site, work came to a halt when a 17-foot reticulated python brazenly slithered onto an active bulldozer. In the rural town of Davie, a 200-pound Burmese python was removed from beneath a storage shed in the backyard of the former mayor. A little farther north, another 200-pound Burmese python was discovered as it struggled to free itself from the chain link fencing of a home in Greenacres. A deceased, 19-foot-long reticulated python was even fished from the middle of Biscayne Bay in a macabre scene at a popular Miami beach. And whether out of avarice or sincerity, highly publicized captures would sometimes spawn dozens of claims of ownership.

Even veteran wildlife trapper Todd Hardwick couldn't imagine the fame that was to befall him on August 17, 1989. A sighting of a large snake beneath a residence in Fort Lauderdale had recently been reported. At the scene, it was confirmed that a hefty reticulated python had wedged itself into an 18-inch crawlspace beneath the home. Over the course of two days, Hardwick set about assembling a team and hatching a plan to remove the freeloading reptile. Among the assortment of tools he had gathered for the job was a large metal prod, which Hardwick used to gingerly coax the serpent from its lair. Alongside a team of six other men, he unearthed a mammoth python which, when finally exposed and stretched taut, measured 20 feet in length and weighed a whopping 250 pounds.

The story of "Big Momma," as she was subsequently christened, became a media sensation. The story caught national attention after being featured by Tom Brokaw on the *NBC Nightly News*. Hardwick decided not to part with the serpent, instead making appearances with his new pet on *Rescue 911* and *The Tonight Show with Johnny Carson*. While the rest of the country marveled at the magnitude of one huge snake, others pondered the effect of the many smaller individuals sure to be on the loose in south Florida.

In 1994—long before pythons were discovered to be reproducing in the Everglades—biologist George Dalrymple offered this prescient council in a report prepared by the Florida Department of Environmental Protection: "The most significant potential problem in the next twenty years has to do with a category of non-indigenous species that is increasingly found as accidental, or intentional, releases in urban and suburban settings: the large constricting snakes (pythons and boas), and large predacious monitor lizards." Though Dalrymple was doubtful of the possibility that egg-laying species like pythons could ever establish themselves, he was the first to publish on the successful breeding of the boa constrictor in Miami-Dade County. He worried that the size, appetite, and longevity of such snakes could pose a threat to native species, domestic animals, or even small children. "Some means of regulating their sale and maintenance appears to be worth exploring," he wrote. "Not everyone is equipped to maintain such large animals." During a recent conversation, biologist Snow praised Dalrymple for attempts at sounding an early warning about the potential for harm: "He got the larger issue correct—he just picked the wrong species."

5
Trial by Fire

In June of 2009, ten Burmese pythons made a lengthy trek from the wilds of the Florida Everglades to the lowlands of South Carolina. They traveled over 600 miles within the dark confines of several tightly sealed Rubbermaid containers during their one-way journey to the Savannah River Ecology Lab, a sprawling research facility operated by the University of Georgia. A small but exuberant team of researchers was present to receive them when they arrived.

Within days, each of the snakes was implanted with a radio transmitter, lodged deep beneath the patterned skin of their backs. Each was also surgically implanted with diminutive data loggers, capable of taking and recording periodic measurements of the animals' body temperature and overall condition. Both devices served to counteract the silent secrecy of the snakes—providing a stream of information on the location and overall condition of each.

Once outfitted with such technological wizardry, the snakes were released into a semi-natural outdoor enclosure specially constructed to house them outside Aiken, South Carolina. Measuring several hundred feet in perimeter and guarded by a smooth-walled fence standing eight feet in height, the escape-proof pit housed the serpents in a mixed environment comprised of live vegetation native to the Southeast, brush piles, underground chambers, and an expansive freshwater pond. All were large males formerly collected from the wilds of south Florida and all were destined to become involuntary subjects in a year-long study designed to test the very limits of their physiology.

Certainly the biggest challenge facing natural resource managers in south Florida is dealing with the unknown. Nowhere else on the planet has a species of python escaped its native range and established a large breeding population elsewhere. The Everglades is ground zero—the only known location where a drama involving giant, exotic snakes is slowly unfolding. What few examples of problem snake management exist elsewhere on the globe deal primarily with significantly smaller species with entirely different life histories. Thus there really are no previous occurrences from which to draw upon past experience, derive inspiration, or find encouragement. Of necessity, control efforts against Burmese pythons began from scratch, built upon the limited information that could be gleaned regarding the habits and movements of these new invaders in the River of Grass.

The novelty of south Florida's python situation, coupled with the singular nature of the problem, garnered interest from around the world—galvanizing attention once again on the issue of exotic species and providing compelling support for legislation aimed at preventing future invasions. Still, while policy makers began debating the practicality of hypothetical regulation, the proliferation of potentially 16-foot-long serpents in an ecosystem being "restored" to the tune of $20 billion was a reality that required more immediate attention.

In an effort to assemble as much expertise on the issue as possible, an invasive snake management and response workshop was convened during July of 2005. The gathering brought together many of the foremost authorities on the biology of snake invasions for an opportunity to brainstorm and exchange information. Ultimately, the event would highlight a variety of glaring information gaps, with participants suggesting means to learn more about when, where, and how the objects of their attention were moving about the landscape. Answering that question, however, would mean following a few around for a while.

Radio telemetry is a common research tool that allows scientists to track the movements of wildlife over time. Animals that are being monitored are fitted with relatively small, unobtrusive transmitters that broadcast a constant string of beeps via radio waves. Armed with a receiver, antenna, and a working knowledge of the frequencies at which the transmitters are broadcasting, a field technician can zero in on the whereabouts of each individual animal with accuracy. The technique has been used with myriad species of concern, and has helped throw the natural history of these organisms into greater relief. Telemetry studies have helped elucidate the home ranges of mountain lions in the American West, exposed secretive chapters in the lives of Atlantic sea turtles, and provided greater detail about the migration of geese over Canadian skies.

Though radio telemetry mandates the use of some specialized gear, decades of use by field researchers around the globe have helped drive down the cost of equipment. At the same time, advances in technology continue to deliver smaller, lighter transmitters, allowing for their use on an ever-expanding diversity of taxa. Radio telemetry is now routinely used to track organisms as tiny and delicate as bats, beetles, and dragonflies. While some scientists are tapping into the increased functionality and convenience offered by the Global Positioning System (GPS), the steep costs of the necessary gadgetry ensure that most cash-strapped research programs will likely continue to practice the cheaper, tried-and-true art of radio telemetry.

Telemetry has been used extensively in monitoring the threatened and endangered biota of the Everglades. Over decades, transmitters have broadcast the locations of Florida panthers, American alligators, and white-tailed deer. Field technicians have spent countless hours donning headphones in pursuit of the elusive beeps emitted by tagged Florida manatees, Everglades snail kites, and white-crowned pigeons. Thus, when faced with the need to learn more about the movements of Burmese pythons in the Everglades marsh, researchers naturally looked to radio telemetry for answers.

Snakes are, however, conspicuously devoid of the necks and appendages where, customarily, one could securely harness an external transmitter. Early efforts to track serpents using telemetry had researchers forcibly feeding transmitters to snakes. Problems with reception, regurgitation, and perceived changes in the snakes' foraging behavior led scientists to begin experimenting with the implantation of transmitters in the mid-1970s. By 1982, Reinert and Cundall had published what is perhaps the most widely used implantation technique today—an involved process of subcutaneous placement near the gonads of the snake, followed by the threading of a whip antenna internally along the length of its body. Mercifully, the paper includes instructions on the appropriate dispensation of anesthesia. In 2006, four Burmese pythons captured in Everglades National Park would undergo a similar procedure and subsequently shed new light on the mysterious workings of an otherwise cryptic invader.

In June of 2006, a 15-foot female python took up residence just outside the eastern boundary of Everglades National Park, where she remained relatively motionless for weeks. Following her prolonged idleness, she channeled her energies into playing the role of a reptilian Easter bunny and, on the 17th, quietly deposited 14 infertile eggs (not an uncommon occurrence among snakes) in a nest atop a lonely embankment on the outskirts of Miami. Still, no amount of camouflage or secrecy could shield the event from the prying eyes of the scientists who arrived on scene shortly after. This, after all, was exactly what they had been hoping for when they christened her "Python 6" in the lab only weeks before.

After being captured along the main park road near a heavily trafficked visitor area, Python 6 was spared euthanasia in favor of becoming one of the first pythons to be tracked through radio telemetry. Though done in the name of science, the prospect of releasing large, potentially fertile pythons back into the wild following hard-won captures was a concern for both management and the scientists themselves. In the past, transmitters have been known to fail, and no one wanted to shoulder the blame for losing even one python to the marsh. To allay their concerns, researchers opted to implant two transmitters in all four snakes involved in the initial study (Figure 10).

In late May 2006, the four pythons were released into a state-owned retention area just outside the eastern border of Everglades National Park where, twice weekly, researchers would track them on foot. To avoid disrupting their natural movements, researchers used triangulation to remotely pinpoint the location of the snakes, limiting visual location to only once every two weeks. If the snakes were to migrate beyond the general area, researchers could alternately track the serpents from a small plane overhead.

During the first two months, little notable movement was recorded for the snakes. Python 6, having laid her clutch of unviable eggs, was removed from the study shortly after due to persistent difficulties with her transmitters. As the summer months marched along, daily subtropical rains worked to raise the water level within the retention area. The deepening waters of the marsh seemed to incite in each of the remaining pythons a desire to indulge their semi-aquatic instincts. Riding the tide of rising waters, the pythons began a rapid series of movements that would take them over extraordinary distances. During the next four weeks, aerial surveys would reveal that one of the pythons, a male, traversed 14 miles of open freshwater prairie to its original area of capture. Another male moved nearly 20 miles in the general direction of where it had been first found, before trekking an additional 14 miles northwest to settle in the Big Cypress Swamp. And Python 5, the sole remaining female, traveled nearly 22 miles over the course of 75 days before settling in the remote area of the park where she was first discovered.

The results of this study were notable for several reasons. Never before had such an exhibition of "site fidelity"—an organism's proclivity to return to a given home range—been documented for Burmese pythons. Similarly, the surprising rates of travel and distances covered while motivated by this homing instinct did not appear in other accounts of the species.

The researchers also considered what the results could mean for the dispersal of pythons elsewhere. In addition to site fidelity, a suite of additional factors could similarly motivate migration over long distances, including mating behavior, natural disturbance, displacement by competitors, or the dispersal of juveniles. Any or all of these could encourage the colonization of pythons into new areas beyond their current range, particularly if these new areas were connected by water. Indeed, by promising to bring wetter hydrologic conditions to the parched River of Grass, planned restoration efforts might also provide new avenues to assist the dispersal of undesirable aquatic pests.

Over the next few years, the rate at which pythons were captured in and around Everglades National Park continued to escalate with the trajectory of previous years (Figure 11). More troubling, however, was the realization that pythons not only could, but were, spreading beyond the confines of the park. Pythons had already appeared on the island of Key Largo in the Florida Keys, but they were now starting to be found north of the park with regularity.

Several pythons were found in the midst of the brackish-water mangrove forests of Rookery Bay on the southwest Florida coast. Only one week later, a nine-foot python was found in the ritzy enclave of Marco Island just five miles south—fueling speculation that it may have arrived in a landscaping truck or found passage over the S.S. Jolley Bridge that links the island to the mainland. Pythons were being encountered with greater regularity along the Tamiami Trail and the levees of the water conservation areas to the north, and reports from Big Cypress National Preserve became more frequent. To the east, large constrictors were appearing along canal banks and agricultural fields adjacent to the park, probably attracted by rodents, domesticated livestock, and fowl. And, thanks mostly to telemetry and a bit of dumb luck, half a dozen additional nests were also found—most containing viable eggs.

Encounters with Burmese pythons farther north in Florida left many to wonder about the potential origin of these individuals. One angler, for example, caught an 11-foot, 40-pound python while fishing the waters of the Indian River Lagoon. And in Okeechobee County, a 17-foot-long, 200-pound Burmese python was killed after an 11-year-old boy found it slithering along a local canal bank. Both serpents were discovered hundreds of miles from Everglades National Park, and the freakish encounters were largely written off as additional instances of escaped or liberated pets. But given what was now known about the potential overland movements of pythons, it was not beyond the realm of possibility that progeny of the Everglades population, perhaps even from successive generations, were now becoming distant outliers on the south Florida landscape.

Members of the media covering the spread of Burmese pythons beyond Everglades National Park grew increasingly trained on a single question: how many snakes were potentially slithering amidst the sawgrass? Certainly, the desire to quantify the problem was understandable. There is a deep-seated desire in the human psyche to size up that which threatens us. And from a management standpoint, there was a practical use for such information. If there were a reliable means of ascertaining a snapshot of how large the population might be, then over time one could gauge whether or not control efforts were having a negative effect on numbers.

Known densities from a handful of studies conducted in their natural range—coupled with a small subset of data known from a limited area in south Florida—provided park scientists with a crude educated guess of population size in the Everglades. But the resulting estimates admittedly ranged widely: somewhere between 5,500 and 137,000 individuals. Given that hundreds of virtually undetectable pythons were being captured by scientists annually, the low-end estimate appeared far too conservative. And the high end of the spectrum, though in the realm of possibility, likely represented an improbable worst-case scenario. The true number was almost certainly somewhere in between.

But the media machine often pays little mind to scientific nuance. Wanting to grab the attention of viewers and readers, outlets rabidly published and broadcast the alarming, high-end estimate without qualification. In a classic manifestation of a schoolyard telephone game, published numbers began escalating to 150,000, then 180,000, and even 200,000. Critics rightly began questioning the figures. "I'd like to know how they come up with that stupid exaggeration," said Vernon Yates, founder of a rehabilitation facility in Seminole. "I believe it's probably around 1,000." But for all his personal experience, Yates' guess proved nothing more than wishful thinking—over the years, park staff had already removed over 1,300 pythons from the marsh.

Python population estimates in south Florida had been developed and subjected to debate. The movements of pythons were being tracked, yielding new insights and allowing for the discovery of nests. Necropsies continued unabated, diets were being documented, and sightings were being compiled. As researchers cultivated an ever-greater appreciation for the habits and potential impacts of Burmese pythons in the New World, one inescapable fact became increasingly clear—how difficult they would be to control.

Both the hunting and reptile hobbyist communities of south Florida repeatedly made offers to assist with management. The former, confident in their ability to point and shoot, implored the park and nearby management areas to allow them access—arguing that their collective experience was the state's best option in the war against wayward snakes. Reptile enthusiasts, meanwhile, encouraged land management agencies to bend long-standing prohibitions on collection in the name of invasive species management.

But ample experience with pythons in the field illuminated just how cryptic they could be. Armed with earth-tone coloration and a penchant for submersion, the animals seemed perfectly suited for a life of hidden seclusion in the Everglades. Most days, one would have had a better chance of finding the Skunk Ape—the foul-smelling cousin of Bigfoot that a handful of locals claim haunts the Glades. Even armed with expensive transmitters and years of field experience, researchers repeatedly encountered great difficulty visually locating their subjects thanks to a complete lack of contrast between patterned reptilian skin and a busy background of vegetation. And looking for signs of movement provided little help—the snakes were fully capable of moving through the marsh with nary a rustle of grass nor a ripple upon the water's surface. Following several years of tracking, field staff were well aware of the challenges and limitations of visual searching. Hunters and hobbyists may have found purchase along roads and levees, but they would be thoroughly outmatched in the Everglades wilderness.

In 2003, researchers at the University of Florida investigated the practicality of using unmanned aerial vehicles (UAVs) for wildlife surveys. The study provided cautious optimism on the ability of UAVs to offset the costs and dangers associated with the use of manned aircraft in aerial reconnaissance. In their never-ending search for new tools, researchers in south Florida questioned whether or not UAVs, fitted with heat-sensing cameras, might provide new hope for surveying the habitats of the Everglades for pythons.

Not surprisingly, Burmese pythons are encountered most commonly along the roads, trails, and levees frequented by people in south Florida. Finding those that persist in the vast acreages of water and wilderness, however, poses a far greater challenge. When mounted to aircraft overhead, infrared cameras could theoretically extend the limits of visible wavelength and allow for the detection of pythons hidden in the marsh. This would only be possible, however, if the animal's body temperature was considerably different from that of its background environment—a relative rarity for cold-blooded organisms. Stars could align, however, during certain times of the day and seasons of the year, depending on landscape variables and animal movement.

Given the narrow window of conditions seemingly necessary, the National Park Service and the United States Army Corps of Engineers organized a two-day workshop to conduct field tests and analyze results. During September of 2009, representatives from several UAV and thermal camera vendors converged in Everglades National Park to showcase their wares. High atop both a telescoping boom and bucket truck, different models of infrared cameras were trained over an open prairie along one of the park's roads. Amidst tall grasses swaying gently in the breeze, two wranglers overturned a large, black Rubbermaid container, releasing a slow-moving python that disappeared almost immediately into the landscape. For the balance of the morning, and well into the late afternoon, thermal cameras recorded an endless stream of images attempting to capture a telltale serpentine shape in the vegetation.

The group reconvened the following day to review the trial, and the results were intriguing. Several cameras, particularly during afternoon hours, had captured striking images of warm snakes popping out amidst dropping temperatures brought about by an unexpected rain storm. Optimism, however, was tempered by the realities of use in the field. After all, pythons were not likely to be found in such an exposed manner in the wild, but rather, would likely take cover beneath a shroud of taller vegetation. Differentiating smaller pythons from the gamut of native snakes could also prove difficult given the resulting monochromatic images. The very use of UAVs in the park could be complicated considerably thanks to Federal Aviation Administration regulations and restrictions on flying aircraft in Congressionally designated wilderness areas. And presuming all hurdles could be surmounted and the technology successfully used to locate pythons in the heart of the Everglades backcountry, how would responders reach remote individuals in time to make a capture?

Recognizing the inherent difficulties with using UAVs, the team conducted a second field trial the following November using an infrared camera mounted on a helicopter contracted by the park. The aircraft flew low over a roughly 40-acre farm field adjacent to the park presumed to contain pythons. A tractor slowly disked the field, plowing under a vegetative mat standing roughly a foot off the ground. By the end of the day, researchers following behind the tractor on foot had documented the presence of six pythons—not a single one of which was detected through the airborne imaging device. To even the most advanced technological wizardry, the snakes too frequently assumed a cloak of invisibility within the welcoming Everglades.

With a better understanding of how Burmese pythons were surviving on the south Florida landscape, researchers began to consider a much larger question. Given what was now known about the snakes' movements, habits, and prey—exactly how widespread would their ultimate impact be in the New World? Was south Florida destined to carry the burden alone, or could pythons spread far and wide, infiltrating lands well beyond just the Sunshine State?

In February of 2008, several researchers with the United States Geological Survey published a paper that seemed to shed some light. The team compiled the mean monthly data for temperature and rainfall from 149 points along the boundary of the native range of the Indian python (*Python molurus*). The goal was to identify the climatic extremes at which the species occurred naturally. This information could then be compared against known climatic conditions in the United States to determine how far, theoretically, the introduced population could spread.

In their paper, the researchers laid bare the errors of our popular portrayal of pythons as inhabiting only mist-filled, tropical jungles. Rather, save for exceptionally arid environments, the Indian python was frequently encountered in a range of far more temperate climes. The snakes were known to persist in areas that experienced average monthly temperatures ranging from a balmy 98° Fahrenheit to a chilling 35°, even tolerating landscapes that grew parched from little to no rainfall over two consecutive months.

Armed with the known constraints of temperature and precipitation, the team was then able to create a climate suitability map of the United States (Figure 12). The results were eye-popping. Based on these criteria alone, it appeared pythons were capable of colonizing the entire lower one-third of the nation. Indeed, it appeared the snakes would feel equally at home in nearly all parts of the Deep South as they did in the Everglades. In fact, they had the potential to go bi-coastal—their potential range extending beyond Texas into the arid southwest and occupying most of coastal California. "According to 2000 census figures," the authors wrote, "about 120 million Americans live in counties having climate similar to that found in the native range of the python."

The researchers took the investigation one step further. The year prior, the Intergovernmental Panel on Climate Change had released its fourth synthesis report—which predicted a likely rise in global temperatures of several degrees over the next hundred years. Using climate models developed by the National Center for Atmospheric Research, the USGS team modeled how the suitability map might differ by 2100. Even assuming the most conservative estimates of warming for the coming century, climate suitability changed considerably—potentially bringing pythons as far north as our nation's capital (Figure 13).

In their published paper, the researchers were quick to clearly disassociate their climate suitability maps from predictions of expansion. After all, the study was based upon only two parameters out of a suite of environmental factors that could promote or prohibit the spread of an organism. The scientists pointed to prey availability, population genetics, and reproductive failure as additional factors to be considered, and noted that our understanding of such variables is poor even in the pythons' native range. Yet such rational, level-headed caveats seemed wholly unnecessary, as they would be ignored by both the media and critics alike.

Concern over impacts resulting from the potential spread of the Burmese python began to take root beyond just south Florida. Farther up the state, in coastal Sarasota County, academics proclaimed the Burmese python a greater immediate threat than sea level rise. Authorities in Texas and Mississippi, concerned their favorable climate might also prove conducive to invasion, began pushing for permit systems to better regulate the ownership of large constrictors. The Burmese python even made one watch list in the United Kingdom, which rated them a high risk species of concern when coupled with a warming climate.

In the face of mounting scrutiny and under threat of tightening regulations, reptile hobbyists and business owners within the footprint of the USGS projections began offering criticisms of the study. For the most part, opinions were largely rooted in personal experience. "I just don't think the projections for them getting this far in the wild in any numbers are accurate because of the cold snaps we tend to have in winter," one Alabama enthusiast told *The Gadsden Times*, citing the cold-induced death of one of his own captive ball pythons.

Without question, David and Tracy Barker were the most vocal of all critics. In an article printed in the *Bulletin of the Chicago Herpetological Society*, the pair sternly called into question the validity of the USGS climate suitability study. Citing the work's failure to acknowledge differences in both taxonomy and genetic variation, the Barkers lambasted the study as an instance of "ecoterrorism" and a perceived "self-serving attempt by federal biologists to bully and intimidate the American public into supporting unnecessary regulation, research, and grants."

The Barkers are an accomplished husband-and-wife team that boasts decades of shared experience in reptile husbandry—particularly pythons. Together they have published extensively on the life history and care of pythons and have authored several definitive works on members of the genus. The couple earned fame among enthusiasts in the 1990s for maintaining one of the largest and most diverse collections of pythons in the world. Over the years, they have come to enjoy considerable respect among breeders, dealers, and keepers for their prolific writing and wealth of experience.

The Barkers are the sole proprietors of Vida Preciosa International (VPI), a company based in the rugged hills 30 miles north of San Antonio, Texas, that has specialized in the rearing and sale of pythons since 1990. Over the years, the Barkers have authored a variety of opinion papers that weigh in on issues related to south Florida's population of problem pythons in a variety of publications. They penned an open letter to park scientists asserting it was equally plausible to blame biologists and animal rights activists for the introduction of Burmese pythons into the Everglades as it was to pin it on irresponsible pet owners. They published a four-page indictment against the idea that the snake trade should be subject to regulation as a precautionary measure against future invasions. And during this time, they have engaged in a prolonged back-and-forth with researchers, agencies, and administrators on the validity of the USGS climate suitability study.

Though perhaps the most vociferous, the Barkers were certainly not alone in their criticisms. In August of 2008, only months after the highly publicized release of the USGS report, a team of researchers published the results of a study suggesting that ecological niche modeling strongly contradicted earlier estimates of potential northward expansion. In their paper, scientists from The City University of New York represented the earlier study as having predicted "that a significant portion of the continental United States provides ecological conditions suitable for snakes." In their subsequent investigation, the researchers incorporated temperature and rainfall extremes and seasonal variability—important additions to the monthly averages examined by the USGS. The result was a distribution map that surprisingly prognosticated no northward expansion in the U.S. beyond where the serpents were currently found (Figure 14). Furthermore, attempting to mirror the 100-year climate projections used by the USGS resulted in a *contraction* of available habitat in most areas of the New World—save for a curious swath of newly hospitable habitat in the Pacific Northwest (Figure 15). Like the waters through which they might make their predicted migrations northward, the truth behind the potential impact of Burmese pythons in the United States was growing increasingly murky.

Lost within all the bickering regarding the plausibility of northern expansion were concerns raised initially in the USGS climate suitability study. While pythons from the Everglades may or may not be able to unfurl like a living, contiguous mat across the country, what about those places with truly hospitable climates? Might places like Texas and southern California still be at risk of invasion by new populations thanks to more accidental and/or intentional releases of large snakes? The question transcends mere philosophy. Pythons have already been previously captured on the loose in both areas—often showing little to no sign of distress. And both areas have historically been prone to natural disasters that could precipitate mass releases similar to those observed in south Florida. And slipping off the blinders of nationalism, what are the implications for our neighbors below the border? Could the coastal forests and humid jungles of Mexico become the next conquest of a spreading population of pythons?

Amidst the cacophony of dissenting opinion, whom was the general public—and its elected representatives—to believe? Officials from the USGS who routinely claim impartiality and disavow themselves of politics? Amateur herpetologists spouting claims supported by their sometimes limited personal experience? Respected authorities sporting heavy business interests in scuttling regulation? Or academics sporting differing methodologies and competing results?

By fall of 2009, the ten Burmese pythons from the Florida Everglades had settled nicely into their new digs at the Savannah River Ecology Lab. From the outset, they ate well and readily occupied the various natural features that had been constructed for their use. They seemed to genuinely find their new environment comfortable—so much so that they made no appreciable attempts at escape. And there was little incentive to do so—they had thus far enjoyed a steady supply of food and the warmth of a South Carolina summer.

But in bringing them to this outdoor laboratory, Michael Dorcas, a researcher with Davidson College, had not intended to merely provide the pythons with a comfortable retirement. Instead, he and his colleagues were preparing to test the physical limits of the snakes to better support or negate conflicting claims of potential range expansion. If they couldn't survive a winter in the Deep South, they probably wouldn't stand a chance in Washington, D.C. But, if a few managed to survive the assails of wind and snow, then there might be reason to fear a steady march north. The ten snakes were to bear a heavy burden upon their patterned backs—made to suffer the coming months of freezing temperatures as a proxy for the entire population of Burmese pythons in the Everglades. And already, the leaves were beginning to turn.

6
Cold Blooded

Early on the first day of July in 2009, I found myself standing before a large audience at the University of Miami Rosenstiel School of Marine and Atmospheric Science. In an intimate seminar room overlooking the flat, blue-green waters of Biscayne Bay, I addressed fifty participants from a popular high school summer enrichment program. Exotic species were the topic of the day and Burmese pythons, of course, figured prominently in the conversation.

The students listened eagerly as, slide by slide, we unraveled the decades-long story behind the arrival, establishment, habits, and control of these rogue reptiles. Two large projection screens at the front of the room were the stages across which scrolled a disturbing pictorial. Over the course of an hour, eyes grew ever wider as I presented image after image of hefty pythons found in the Everglades marsh. In one was an alligator consuming a python—while another clearly showed the tables turned. I later projected images showing a captured snake stretched lengthwise across a laboratory table for the purposes of implanting a radio transmitter, the serpent held fast by a dozen strained, human arms. Later images showed the innards of several pythons laid bare by the sharp scalpel of a scientist post mortem. From some, stomach contents were removed and photographed to document prey. From others, dozens of eggs were unearthed from the confines of their body cavities—undeveloped young which might have one day added their numbers to the population. Through it all, the students remained transfixed, captivated at the situation that had befallen both the Everglades and the pythons.

As my allotted time came to a close, I offered what I had intended to be a rhetorical question. "How many people do you know," I offered, "who could properly care for a sixteen-foot-long two-hundred-pound serpent?" An expected absence of hands filled the air. "Well, right now," I continued, "anyone—and I mean *anyone*—can legally own any one of the largest snake species on the planet for the cost of an iPod. It's important for us as a society to ask ourselves whether or not that makes sense." Though I hadn't asked for an answer, they replied nonetheless with a tone of considerable collective disdain: "No way, man."

I concluded my presentation at 10:30 A.M. In retrospect, one could contend that the context of my presentation, and the leading manner of my questioning, may have significantly biased the opinions of my young and impressionable students. Yet at the very moment I was delivering my program that morning, a tragedy was unfolding in central Florida that could only steel such sentiments in the minds of those who would hear of it in the days and weeks to come.

By 10:30 a.m., Sumter County paramedics had already arrived at the house of Shaianna Hare, only to find the blue-eyed two-year-old lifeless. The only trauma visible on her tiny body was a tell-tale arc of punctures atop her head—perfectly matching the teeth of a mutinous eight-and-a-half-foot family pet. Charles Jason Darnell, the boyfriend of Shaianna's mother, awoke that morning to find their albino Burmese python, Gypsy, missing from its glass terrarium. Sometime during the night, guided by a keen sense of smell and hunger pangs, the reptilian predator entered Shaianna's room and ascended her crib. It was there that Darnell eventually found the predator—embracing the girl in a series of strong, wicked coils that had long since robbed her of breath. In his panic, he stabbed the serpent in a desperate attempt to pry the creature's tense muscles off its victim but, as evidenced by replays of his subsequent 9-1-1 call, he was keenly aware that the girl had already expired.

When authorities arrived, they found Gypsy loose in the family's trailer, still bleeding where Darnell had awkwardly plunged his blade. The snake was quickly captured, bagged, and whisked away to a caretaker pending an examination to determine whether or not the snake should be euthanized. At the time, the State of Florida listed Burmese pythons as *Reptiles of Concern*, requiring a special permit for their ownership, sale, or exhibition. As it turns out neither Darnell nor Jaren Hare, Shaianna's mother, was properly licensed to keep their hefty pet. Had there not been fatal implications, they might have only faced a second-degree misdemeanor. Instead, both individuals were charged with third-degree murder, manslaughter, and child neglect.

Two years later, a jury of their peers would find Darnell and Hare guilty of all charges, following three days of testimony during which the prosecution attempted to illustrate the manner in which the pair had endangered their daughter. Failing to appreciate the latent instincts present in even the tamest constrictor, the pair exhibited little concern for properly caring for or securing the serpent. At the time Shaianna was fatally attacked, the snake was being kept in its glass terrarium with a lid made of nothing more than a safety-pinned quilt. Additional testimony detailed how the pair indirectly put the toddler in harm's way. Experts testified that while a healthy eight-and-a-half-foot-long captive python might weigh around 150 pounds, Gypsy tipped the scales at only 13 pounds at the time of the attack. Chronically undernourished, the snake was likely motivated by desperate hunger. Unable to afford food for the animal, Darnell had fed the snake nothing but a single road-killed squirrel he had found a month before.

Fatal encounters between innocent children and predatory wildlife always garner intense media attention. These incidents usually elicit public sympathies when they are the unintended consequence of human encounters with animals in wild, natural settings. Venturing into wilderness does, after all, entail certain risks that people can only hope to minimize at best. But when people knowingly bring potentially dangerous animals in close proximity to others, public sympathies can quickly give way to harsh criticism and blame. Injury or death resulting from predators introduced into one's own surroundings elicits little pity for all but the youngest victims, as it seems to flout generally accepted common sense and the lessons learned from collective past experiences.

There are numerous animals whose strength and aggression and potential to harm humans are well documented. Bears, mountain lions, and alligators, for example, have all proven to be formidable predators in their natural settings. Furthermore, each has been implicated in numerous human fatalities. There is a well-founded general sentiment that such animals have repeatedly shown themselves too dangerous for domestication, suited only for the wilds in which their bodies and instincts are forged. Most people view such animals as remarkable works of nature worthy of admiration from a distance—not lethal ornaments intended to grace their living rooms as a conversation piece.

Yet in homes across the United States, large constrictors are often treated as little more than showpieces to be counted among their owners' furniture and other belongings—this despite a lengthy assortment of observations and experiences that help solidify their reputation as potently muscled predators. As has been noted previously, the predatory behaviors of pythons are well documented in science, and it is folly to believe that constrictors halt their aggressions at the threshold of human encounters.

The Hippo Lodge is a rustic outpost just east of Katimo Mulilo in southern Africa. The facility is situated on the banks of the Zambezi—one of the largest rivers on the continent. Here guests can rent canoes for self-guided trips on the river, though would-be adventurers are cautioned to embark at their own risk. While most guests enjoy a day on the water unscathed, Febby Katwa knows well the possible tragedy that could befall the unlucky.

Katwa reported to the Hippo Lodge on February 6, 2001, where he worked tending the sprawling gardens on property. Around midday, he found himself walking through tall grasses along the boundary of the lodge. Suddenly, before he could even begin to comprehend what was happening, weighty coils encircled his body and began applying pressure. What was likely an African rock python had ensnared the 29-year-old gardener by the calf and quickly began the excruciating process of asphyxiation. "It was so strong I immediately smelled death," Katwa would later recount. "I knew I was going to die."

Stunned bystanders watched helplessly as the 16-foot roughly 200-pound serpent continued to crush Katwa within ever-tightening circles. Armed with only a carving knife from the nearby kitchen, the lodge bartender rushed to the scene and severed the python's head. As the dying snake loosened its binding grip, Katwa was extricated and treated for the severe gashes inflicted by his attacker's long, recurved teeth. Weeks later he would reveal the details of his ordeal, and confess to suffering repeated nightmares about the incident.

That Katwa lived to have nightmares proved lucky, for not all encounters have ended so well. In 2003, a 38-year-old woman was attacked and killed by a nearly ten-foot python while collecting firewood in a forested area of southeastern Bangladesh. The snake was beaten to death by a group of villagers before recovering her lifeless body, which had been swallowed to the waist.

The majority of human/python encounters involve "faceless" victims with whom we are rarely familiar. These encounters become far more real, however, when they befall familiar personalities. In 2007, popular nature series host Brady Barr made headlines when the object of his pursuit turned violent. Barr and his crew had been tracking reticulated pythons while wading through waist-deep water in a flooded cave in Indonesia. Unexpectedly, one of the snakes suddenly ripped into the host's right leg with a full armament of sharp teeth—drawing howls from Barr. "I was so completely incapacitated by the pain," Barr later wrote, "I was terrified that the snake was going to pull me off my feet with its coils around my legs and drag me underwater." A surprisingly strong reaction given that the snake with which he tangled was relatively small—measuring only seven feet in length.

Barr would return to the same cave only two months later to capture additional footage. For those in the business of producing documentaries, the final product can certainly be worth the risk. Pythons star in countless features and documentaries precisely because, in their natural range, they prove time and again to be the powerful, opportunistic predators we imagine. Past experience has also illuminated a potential for violent and seemingly arbitrary transgressions against people in the wilds of southeastern Asia or sub-Saharan Africa. At some point in our past, we may have found comfort in knowing that such dangers, though admittedly rare, could only be encountered half a world away. But, as with so much else in our present world, oceans no longer protect us.

———

Though pythons are not native to North America, they are encountered with surprising regularity in our nation's fifty states. Excluding extreme southern Florida, pythons sighted in the United States are presently assumed to be isolated individuals—either recently escaped or released pets. Often, these rogue individuals spawn considerable media coverage, if not for the location in which they are found (storm drains, washing machines, vehicles, and the like), then for the encounter through which they are finally detected. Most serpents are masters of disguise that are supremely capable of "lying low" for long periods while on the lam. They can elude prying eyes, however, only until instincts force a brazen transgression in suburbia.

Being only an eight-pound rat terrier, Max was certainly the underdog when an 11-foot Burmese python showed up at the front door of his home in West Palm Beach, Florida, in 2006. His owner watched in horror as the snake grabbed little Max by the head and began throwing loops around his slight body. "It had at least two coils around him," Wayne Vassello would later testify in a court of law, "and there was a lot more snake left." Using a golf club, Vassello beat the snake into eventual retreat, but not before mortally wounding the dog. Max, badly hurt by the experience, died of his injuries the following morning. A judge would later order the snake's custodian, John Corkan, to pay Max's owners $1,300 in restitution for negligence.

Though preying upon small dogs might not be surprising, pythons seemingly fail to differentiate between rat terriers and pit bulls. In early October of 2001, a man phoned city police in Merced, California, to report the disappearance of both his pet Burmese python *and* his nine-month-old pit bull. By the time authorities arrived, the man had found the snake beneath his house—with a noticeable 30-pound bulge around his midsection. Both mysteries solved.

But in the United States and beyond, where these animals remain in captivity and are domesticated through regular handling, does the threat of fatal encounters with humans not diminish? When these animals are confined securely, fed regularly and afforded the utmost care, do we not null the chances of injury? And when placed in the hands of responsible, knowledgeable owners, does the specter of grave harm not dissipate? Given such assumptions, one might hope the story of two-year-old Shaianna Hare is a unique anomaly. But again, past experience tells us otherwise.

As we have observed elsewhere on the planet, pythons do not afford people any particular measure of regard. We are simply one of countless biological agents with which they are capable of interacting. Most wild pythons—and the same could be said of most snakes around the world—are happy to avoid all contact with humans if given a wide enough berth. But generalities do not promise guarantees, and often, the natural world is fraught with exceptional acts capable of leaving us dazed and dumbfounded. Thus, interactions between people and pythons can arbitrarily range between passive indifference and blatant hostility. They are capable of treating us to violent surprises at any time, and they deliver with surprising regularity.

In April of 2001, a 10-and-a-half-foot African rock python had to be pried off ten-year-old Zack Payne after the serpent unexpectedly sank its fangs into the New Jersey boy's eyelids. The snake, which belonged to Zack's brother, had escaped from its cage and attacked him. "That thing must have grabbed his whole face," said Dr. David Watts, the cosmetic surgeon who treated the boy, noting that puncture wounds dotted his entire countenance.

Such seemingly random attacks actually occur fairly frequently. In 2006, police used a Taser to ward off an eight-foot pet albino Burmese python that had unexpectedly wrapped itself around the left arm of a Uniontown, Pennsylvania, man. Only three weeks later, at the Tarpon Springs Aquarium in Florida, a 14-foot Burmese python plunged its teeth into employee Alison Cobianchi and tried to drag her into its cage. Again, responding officers were forced to Taser the animal when efforts to free Cobianchi's arm and waist by hand were unsuccessful. "It was definitely the most scary and painful thing that has ever happened to me," she later recounted. In 2008, Sergeant Ryan Nelson was hailed as a hero when he saved pet store owner Teresa Rossiter from the coils of a 12-foot Burmese python in Eugene, Oregon. Rossiter pleaded with responding officers to spare the snake (which no doubt represented a significant investment), prompting Nelson to remove the snake from her with only his gloved hands.

Though these stories are both violent and troubling, they do not share a similar ending with that of two-year-old Shaianna Hare. One might be tempted to imagine that the fate that befell the central Florida toddler was a particularly tragic anomaly that has not replicated itself elsewhere in our country. But while Shaianna's story is perhaps the freshest and most poignant in our collective consciousness, in truth, many people have succumbed to the crushing strength of a predatory pet python.

In 2008, a preliminary autopsy revealed that Amanda Ruth Black died of "asphyxiation by neck compression." Investigators didn't have to look far to find her assailant—Diablo was still loose in Black's Virginia Beach home when police arrived. A message scrawled on a nearby dry-erase board indicated the 13-foot-long reticulated python was in need of medication, and the experienced, 25-year-old pet shop employee had apparently only tried to oblige. But, with no one else around to assist her, she was evidently overpowered by the heavy, agitated serpent. Black's husband returned home that evening to find his wife on the floor of their townhome beside the empty, open enclosure where Diablo was kept. When authorities had completed their investigation, he turned the snake over to Virginia Beach Animal Control who, only one week later, euthanized it at his request.

In total, at least twelve fatalities in the United States have been attributed to captive pythons since 1980. In most cases, like that of Amanda Black, the animals unexpectedly turned on their owners with lethal results.

In September of 2006, 23-year-old Patrick Von Allmen was killed in Kentucky by a 14-foot reticulated python his family had acquired only five months prior. Allmen was described as an avid reptile enthusiast who had experience handling reptiles for nearly a decade. In December of that same year, Ted Drees was killed by a 14-foot pet python in his Ohio home. Though impossible to know, it is interesting to speculate about how Drees might have responded had he read about Allmen's unfortunate death only three months prior. Could that precedent have been sufficient to warn him of the dangers posed by large constrictors, or would it have taken additional examples? For there are certainly more to consider.

Brothers Grant and Lamar Williams lived in the Bronx, New York, and aspired to one day become herpetologists. They amassed a growing collection of animals in their apartment, including several milk snakes, garter snakes, and water snakes. In the summer of 1996, the pair paid $300 for a 13-foot Burmese python at a local pet store. Live chickens sustained the 85-pound serpent for roughly three weeks. On October 9, a horrified neighbor found the body of 19-year-old Grant sprawled on the ground in his doorway, bleeding from the mouth, wrapped tightly in the coils of his new pet. "Asphyxia due to compression of the neck by the snake" was listed as the official cause of death upon his arrival at nearby Jacobi Hospital.

Similar cases took place in 1983, when a man was crushed to death by his 16-foot Burmese python in Missouri, as well as in 2002, when 43-year-old Richard Barber was killed in Colorado by Monty—the 11-foot Burmese python he had kept for over five years. Only three years prior, Barber had been cited for keeping a snake larger than six feet against local ordinance, and was ordered to relocate his snake outside the city limits. Monty was euthanized by authorities just one week after Barber's death.

Though such stories are horrific, pity is often in short supply for those that are killed in their attempts to befriend wild and dangerous animals. What makes the story of Shaianna Hare such a bitter pill, then, is how steep a price she paid despite her own innocence in allowing a predator into her home. A lack of proper care around dangerous animals by others can quickly lead to an unfortunate situation for their children. And Shaianna wasn't the first such casualty—or the youngest.

"Hisser" was christened by his owner, Robert Altom, for his less-than-friendly disposition. He and his wife, Melissa, purchased the ornery seven-foot African rock python from a friend for $50. The couple moved the snake into the Centralia, Illinois, mobile home they shared with their three-year-old son, Jessie. The Altoms kept Hisser in a four-foot-long glass terrarium—topped with a wood-framed, chicken-wire lid. Sometime in the early morning hours of August 29, 1999—only three months after acquiring the serpent—Hisser exploited a breach in his enclosure and slithered toward Jessie's sleeping body nearby. Before the boy could cry out, the snake wrapped coils around his tiny frame. Officers arriving at the residence at 9:45 that morning found Jessie already expired. Hisser, meanwhile, was coiled beneath a nearby sofa.

Days after the incident, the state attorney of Clinton County charged the Altoms with felony child endangerment. If convicted, the coupled faced a maximum penalty of ten years in prison. The case brought outcry from those both sympathetic to, and critical of, the Altoms. At issue was whether or not the couple knew the snake could be dangerous. In order to prove their guilt, the state would have to prove the couple knowingly placed the child in harm's way.

In defending his intent to prosecute, State Attorney Henry Bergmann argued, "When you have parents bringing into their home an animal which is obviously capable of killing their child, whether they say they knew is not really an issue. As parents, you have the responsibility to find out." Python expert David Barker appeared as a witness for the defense and testified that after 30 years of experience working with pythons, he would have never predicted such a relatively small snake could smother a three-year-old, leaving the jury to wonder: how could the Altoms have known? He would go on to reveal that he received nothing more for his testimony than a plane ticket, yet felt compelled to act, "because I feel that there could be set a bad legal precedent regarding the keeping of snakes in captivity."

On March 24th, Judge Harold H. Pennock III acquitted the Altoms of the endangerment charge, to the satisfaction of many who believed the loss of their beloved son Jessie was punishment enough for the pair. Ultimately, Judge

Pennock felt the prosecution had not proved its case—that is, had not clearly shown that the Altoms knowingly placed their son in jeopardy. Any number of factors may have contributed to this opinion: the couple's conspicuously sincere sense of loss and remorse, their limited financial and educational backgrounds, or David Barker's expert testimony. It is an unfortunate reality, however, that recognizing the potential danger a "relatively" small serpent poses to a child required only a look back at the not-too-distant past.

In 1993, 15-year-old Derek Romero was crushed to death by his older brother's 11-and-a-half-foot Burmese python. In 1982, an eight-foot python escaped from his enclosure and killed a 21-month-old boy in his crib. In 1980, a seven-month-old baby girl was killed by her father's eight-foot reticulated python. And in 2001, eight-year-old Amber Mountain was asphyxiated by a ten-foot pet Burmese python that had escaped its enclosure. The snake was one of five owned by her family at the time.

As a parent myself, my heart truly aches at the thought of any child suffering the fate of those detailed in these pages. My sincerest sympathies extend to the parents who likely find themselves quietly tortured by the unending flood of vivid images, sounds, smells, and emotions that accompany the memories of their personal tragedies. That many honestly did not understand the violent potential of their pets I will not argue. That many were confident of their handling abilities and the care they exercised with their animals I am most certain. On the eve of tragedy, I imagine everyone considers himself to be a skilled and responsible pet owner.

And yet tragedy strikes—and strikes repeatedly. It is a pattern echoed many times over. For those who have suffered personal loss, the reality of how dangerous a large constrictor can be is now crystal clear. And for the rest of us, exploring the details of their misfortune helps paint a picture that—with each account—grows ever sharper. Albeit infrequently, all large pythons have the potential to violently, and fatally, injure children and adults alike. And while the consistent use of best practices may help lessen the risk considerably, it takes but one opportunity—just one lapse in care—to yield a disastrous result.

The same can be said, of course, for a much larger array of exotic wildlife currently kept in captivity. And an element of risk exists even with domestic animals typically considered to be of minimal concern. For perspective consider that in 2009 alone dog attacks accounted for 32 human fatalities in the United States. In fact, 2009 was not particularly extraordinary considering the annual mortality from dog attacks since 1979 has ranged anywhere between 11 and 33 deaths annually—the majority of them children. Over the past thirty years, more than 500 people have been killed by canine companions.

Weighing the appropriateness of keeping an animal in captivity, therefore, is as much about the specifics of the organism itself as the surroundings into which it is introduced and the capabilities of those who care for it. Given these realities, is it still prudent to declare inviolate the opportunity for *anyone* with a bit of surplus cash—independent of his abilities and regardless of the constraints of his personal affairs—to legally purchase dangerous animals as personal pets? Is it more important to hold as sacred the right of *every* individual to own pit bulls, bears, mountain lions, and pythons at the expense of the safety of their families and the larger community?

It is a fair question and yet one for which an answer is elusive. But it is a concern worthy of introspection and deliberation, and I suspect the precocious students in my summer class may have rightly answered the question for themselves: "No way, man."

7
The Best Hope

The tiny island of Grassy Key has inspired neither a Hemingway novel nor a Jimmy Buffet song. Amidst the chain of islands that make up the famed Florida Keys tract, Grassy Key is rather unremarkable. Visually, it appears very much like its neighbors to the north and the south—flat, scrubby parcels of earth perched ever so precariously above the waterline. On its sparse soils grows an amalgam of hardy trees and vegetation that have evolved to withstand the salty, windswept conditions of a tropical Caribbean isle. Were it not for the signage each municipality has erected to welcome vacationers to their communities, there would be little to differentiate one land mass from another.

Like other islands nearby, Grassy Key boasts a motley collection of residences, boats, bars, and gift shops. It lacks, however, both the visitor amenities available on the popular islands of Key Largo and Islamorada, and the party-til-you-drop cache enjoyed by Key West. Consequently, most visitors simply motor between their different adventures—be it diving on the reef, fishing on the flats, or drinking at Sloppy Joe's—with little thought to the islands they pass along the way. Grassy Key is one of several islands that seemingly suffer a chronic lack of identity.

Grassy Key has, however, contributed in part to the lore of south Florida. Several publications over the years have perpetuated the claim that the island houses buried treasure that has yet to be unearthed—left behind by one of the many infamous pirates alleged to have sought refuge in the Florida Keys. In more recent years, unbeknownst to most of the south Florida community, a decade-long ecological drama has slowly unfolded on the island. The key has, of necessity, become an outdoor laboratory of sorts—ground zero for a bold case study in the control of invasive species. And as it turns out, the lessons learned from this grand experiment may, in fact, prove far more valuable than mere pieces of eight.

People are seldom delighted or inspired by encounters with rats. A quick glimpse of even the smallest rodent is often enough to propel some into panic, often with lethal consequence to the smaller mammal. The appearance of these voracious scavengers in one's home often conjures up concerns about property damage, disease, and unwanted animal droppings. In pursuit of their eradication, there is no end to the amount of time, energy, and money that is willingly spent on traps, baits, poisons, and exterminators. And yet, others have elected to intentionally purchase some of the largest rodents on earth—as pets.

Rodents occupy the largest order of mammals on the planet, and many of the more than 2,000 species recognized have now been domesticated. Most of us are familiar with the common availability of gerbils, hamsters, and guinea pigs for purchase at pet shops. Yet the desire to obtain and own ever-more exotic creatures has also led to a market for their larger relatives, like South American chinchillas or 140-pound capybaras.

Sometime in 1999, a handful of pudgy Gambian giant pouched rats (*Cricetomys gambianus*) made a daring escape into the wild from an exotic pet breeding facility on Grassy Key. In the years that followed, the initial escapees established a small, reproducing population on the island—even extending their range onto nearby Crawl Key. At the time, Gambian giant pouched rats were becoming increasingly popular in the exotic animal market, and were being domesticated in homes throughout the nation. The same charm and charisma that proved attractive to keepers also served to endear the six-pound mammals to residents on Grassy Key. Though aware of their presence for years, the island dwellers seemed to pay them little mind. The rodents flourished under such *laissez-faire*, growing more numerous and increasingly tolerant of humans and their environs. From time to time, large groups of rats would be found feasting on bowls of cat food placed outside, while the intended helpless recipient looked on sheepishly.

In 2003, an infected shipment of Gambian giant pouched rats in Illinois was implicated in the first known outbreak of monkeypox in the United States. In only a few short months, 79 people would eventually be diagnosed with the disease, spanning six Midwestern states. In an effort to prevent further infections, the Centers for Disease Control and Prevention and the Food and Drug Administration enacted an emergency federal ban on the importation of six different African rodents. Casting a spotlight on the role of rodents as vectors for disease, the story caught the eye of audiences nationwide—including, perhaps, those living in the Florida Keys.

Shortly thereafter, a resident on Grassy Key alerted state wildlife authorities to frequent sightings of giant pouched rats on the island. Citing its now-known potential as a carrier of disease, hefty size, voracious appetite, and reproductive vigor, authorities deemed the foreign species a dangerous threat to Florida's ecology. In relatively short order, a small team of biologists embarked on an ambitious mission of complete eradication.

Following confirmation of the population on Grassy Key, the team began experimenting with a series of detection, monitoring, and control strategies. Through trial and error, researchers tinkered with motion-triggered cameras, toyed with different formulations of baits, evaluated the efficacy of various trap designs, conducted surveys along roads and other suspected avenues of spread, and launched a public awareness campaign among residents. It was quickly discovered that the rats only occupied a specific habitat on the islands—the drier, upland hammocks that also housed the island's neighborhoods. Radio telemetry revealed that individuals could travel over half a mile in a single night, but most were only likely to journey between their burrows and a food source. It was learned that a small dose of peanut butter, oats, and 2% zinc phosphide could kill a rat in less than two days—and the toxic mixture was consumed greedily. And happily, it was determined that the population had not yet spread beyond Grassy and Crawl Keys.

No one knew exactly how many Gambian giant pouched rats occupied the islands, but during the years since their discovery approximately 200 individuals had been trapped. It was clear the animals were thriving and multiplying—a surprising show of resiliency considering Hurricane Wilma had completely blanketed the islands in storm surge in 2005. Between 2006 and 2007, following two years of study, state and federal officials launched an ambitious campaign to eradicate the population from the islands. Securing the permission of nearly 500 private landowners, agents deployed hundreds of live traps and bait stations around both islands. Using a "moving front" strategy, the stations were baited in stages—first using an innocuous peanut butter and oats mixture before introducing the fatal toxicant several days later.

In the weeks and months that followed, few pouched rats were reported dead. In fact, few were reported at all. Most authorities believe the rats, having consumed their final deadly meal, retreated to their burrows to suffer an agonizing death. By June of 2007, federal agents began removing bait stations from the islands, and initiated a monitoring program to detect any remaining individuals. By September of 2010, the state officially declared the noxious pest eradicated.

The saga of the Gambian giant pouched rat on Grassy Key provided perhaps the best opportunity to date for the successful eradication of an invasive species. Early reporting allowed for the containment of the problem before spreading beyond all hope for control. Broad support from a majority of island residents allowed for a near island-wide control strategy. Physical isolation of the rodent population on island ecosystems helped discourage their emigration elsewhere. The overall size and activity patterns of the rats lent themselves to effective monitoring and trapping programs. And thanks to a voracious appetite, the success of ingestible baits as a viable control method was all but guaranteed.

Still, in their analysis of the eradication effort, the biologists were quick to point out that one of the most important keys for successful eradication was minimizing "the risk of immigration." Thanks largely to health concerns elsewhere in the nation, an emergency ban on further importation and sale of several African rats was imposed by the federal government in 2003. The State of Florida also belatedly enacted prohibitions on ownership of Gambian giant pouched rats in 2007. Had legal mechanisms not been used to help close the spigot on new imports and reduce the possibility of new introductions elsewhere, eradication might never have been possible.

In south Florida alone, nonnative plants and animals that have proven themselves capable of invading natural landscapes number over 300—and counting. Every year, like clockwork, resource managers can anticipate the arrival of at least one or two more. Given the limited amount of resources available to combat the issue, most agencies use a simple formula to prioritize how to spend their time, money, and manpower. Those invaders that pose the greatest potential for harm and present the most realistic chance of control will receive the lion's share of attention. Conversely, those that present little risk, or are so widespread that control is perceived to be impossible, often go unaddressed. And somewhere in the middle lie the vast majority of exotic species—infestations that can only be managed as opportunities allow.

The costs incurred in the prolonged tangle with invasive species are inherently difficult to estimate. Some have quoted annual losses resulting from these species as ranging between $100 and $137 billion in the United States alone. Multitudes of government agencies, private corporations, and nonprofit organizations employ thousands of workers in all-out campaigns against invading species. Hand saws, helmets, goggles, gloves, clippers, chemicals, cameras, and crews are loaded upon ATVs, airboats, helicopters, earth movers, electro-fishing boats, and bucket trucks in a large-scale, expensive effort to ferret out landscape interlopers. And though it is possible to estimate losses to industry and the expense of control, costs to human health, productivity, and ecosystem function are nearly impossible to calculate.

The battle against invasive species in south Florida has arguably been waged over the last 40 years. Looking back, then, what has been accomplished over those four decades? By our force and will, how many of these invaders have we fully eradicated from our midst, never to cause havoc again upon our natural landscapes? Presuming even the most generous interpretation of "eradication," the answer is five.

In 1966, several giant African snails were brought to Miami by a boy returning from a trip to Hawaii. The snails, which had proven globally to be highly invasive crop pests when introduced into favorable environments, were released into a garden in North Miami. As members of a hermaphroditic species capable of producing up to 1,200 eggs per year, the population exploded. State agricultural agents were made aware of the growing infestation in 1969 and quickly launched a full-scale eradication effort. Over the next five and a half years, over 25,000 parcels of land were chemically treated with 128 tons of arsenate-metaldehyde at a cost of $700,000. The project consumed over 67,000 hours of effort. By 1975, over 18,000 snails had been destroyed and, following two years with no reported sightings, the animal was declared eradicated.

For several years at the turn of the century, a large population of black-tailed jackrabbits (*Lepus californicus*) darted swiftly about the runways of Miami International Airport. Some, of course, proved faster than others, leaving the more sluggish to occasionally fall victim to speeding aircraft and support vehicles. Though the rabbits themselves posed little threat to the safety of passengers, their rotting carcasses attracted flocks of vultures, creating a more substantial hazard to aircraft engines. In 2003, under pressure from the Federal Aviation Administration, a multi-agency eradication effort ensued under heavy protest from animal rights activists. In a two-part strategy, live trapping of the animals would be tried first. Any that remained would be exterminated. At the end of the effort, 301 rabbits were captured, and an additional 172 were shot on site.

The winds of Hurricane Andrew have been blamed for unleashing far more than just Burmese pythons. The 1992 tempest has also been implicated in the release of numerous sacred ibis (*Threskiornis aethiopicus*) from zoos and private bird collections. One of the few wading birds that readily prey on eggs and nestlings, the appearance of sacred ibis in several rookeries in the Everglades elicited great concern from resource managers. In 2008, a multi-agency control effort was initiated, spearheaded by the United States Department of Agriculture. By late 2009, approximately 75 birds had been removed from a handful of locations in south Florida. Though officials remain reluctant to call the eradication effort successful, the prolonged lack of verified observations suggests this may be another rare management victory.

In October of 2009, 14-year-old angler Jake Duchene pulled an unusual catch from a four-acre retention pond he frequented near his home in Palm Beach County. Unlike the bluegills and largemouth bass he commonly battled, this fish was round, chubby, and sported rows of sharp, pointed teeth. Biologists with the Florida Fish and Wildlife Conservation Commission identified the catch as a red-bellied piranha—an Amazonian predator with an infamous (and often exaggerated) reputation. Duchene's capture marked the eighth time a rogue piranha had been found in Florida waters. A subsequent survey of the same retention pond netted another just ten days later. Unsure if more could be lurking beneath the surface, state biologists laced the pond the following month with rotenone, a short-lived toxicant that quickly enters the gills of exposed fish with lethal results. Rotenone is an indiscriminate control method—killing both target and non-target species within range. Having poisoned every living thing within the pond, biologists would spend the next several days picking through a decomposing carpet of floating remains before finding a third piranha floating among the stinking mass. All three piranhas were adults and there was no clear evidence of reproduction. By technical definition, it did not appear a population had yet established. But given the potential for harm, biologists seized the opportunity to attempt complete eradication—even at the costly expense of the innocent. Months later, state biologists would return to the area to stock the pond with native bluegills and largemouth bass.

Reasons abound why some species are easier to exterminate than others. Often it is a function of physical structure or known behaviors of the organism itself. The size and nature of the area over which the invasion occurs and the size of the overall population can also play an important role. And the availability, or lack thereof, of management tools suited to particular taxa can either help or inhibit the extent to which species can be extracted from a foreign range. Giant African snails are slow-moving, conspicuous, and can be collected and destroyed without great difficulty or public outcry. Gambian giant pouched rats are relatively easy to corral into traps and readily ingest toxic baits, as they feed often and greedily along predictable routes of travel. Black-tailed jackrabbits and sacred ibis are easily spotted from a distance, and their forms present an easy target in the crosshairs of a skilled marksman. And the limited range and number of red-bellied piranhas found in West Palm Beach afforded the nearest possible real-world equivalent of shooting fish in a barrel. Total eradication is possible only when the many variables of science, species, population, and place are properly aligned.

Though the specifics of each story are entirely unique, there exists a common thread that runs throughout each of the five case studies of successful eradication in south Florida. In each instance, management authorities were made aware of the presence of new nonnative species fairly early in the invasion process. As a result, populations generally did not extend beyond more than a small geographic area, and individuals had not reached densities that exceeded all hope for control. Furthermore, confirmation of invading populations was promptly followed by swift, decisive management efforts that were clearly directed towards the complete and utter demise of the invading organisms. After decades of research and attempted control, it has become evident that there is little hope of eradicating the vast majority of nonnative species that become established. But experience has equally demonstrated that early detection, followed by an immediate and unwavering control program, represents the best possible chance.

Given the limited success in eradicating other invasive organisms in the area, the prospect of ever removing Burmese pythons from the Florida Everglades seemed daunting. Unlike the limited geographic range of the Gambian giant pouched rat, pythons roamed freely over hundreds of thousands of acres of wilderness. Unlike the relatively miniscule population of black-tailed jackrabbits, pythons potentially numbered in the tens of thousands, if not more. Unlike the sacred ibis, whose flamboyant coloration made it easy to shoot, pythons used a cloak of camouflage to blend seamlessly into the watery landscape. And while giant African snails were slow and easy to grasp, pythons seemed capable of eluding all attempts at capture.

One particular fact served up an additional helping of despair. Across the broad sphere of our planet, where people daily interact with, study, alter, manage, and otherwise exert their dominion over every known species, not a single invasive reptile has ever been successfully eradicated through management efforts. Some, after having been established for decades, have suddenly ceased to occur. But where this happened, their enigmatic disappearance was precipitated by a curious function of nature and not the result of intentional human intervention. In general, coordinated efforts to exterminate nonnative reptiles are not common. But where they do occur, those efforts have garnered the worst possible success rate for eradication—zero percent. At present, not a single case study exists from which to draw inspiration or precedent. The eradication of Burmese pythons was not only a faint, dimming hope in south Florida—history paints it an entirely unrealistic impossibility.

Burmese pythons did, however, share one thing in common with all exotic species—a blatant disregard for political borders. The snakes continued to move unfettered beyond the boundaries of Everglades National Park, which lies at the southern end of a continuous patchwork of state, county, tribal, nonprofit, and private lands. To the east, they occasionally ventured over the shallow bays and creeks of Florida Bay onto busy US1, where passing motorists unexpectedly encountered living speed bumps stretched across the road as they sped to and from the Florida Keys. In agricultural fields to the east, growers began churning up large pythons as they tilled their fields for the season. And the Miami-Dade Fire Rescue department experienced an uptick in calls from residents requesting the removal of large snakes making unwelcome appearances on rural properties in their jurisdiction.

To the north, the pythons were advancing beyond Everglades National Park into other protected natural areas. Park rangers at nearby Fakahatchee Strand Preserve and Collier-Seminole State Parks began finding pythons amidst the freshwater marshes and mangrove coastlines within their boundaries. Sportsmen in the Big Cypress National Preserve were now running into unexpected, reptilian game at their favorite hunting sites. And along the Tamiami Trail, pythons were regularly found dead along the highway—casualties of passing motorists. At the time of death, some were clearly headed north; others were moving south. It was clear that some snakes had successfully dodged traffic and were now occupying the roadless, watery expanse of the Water Conservation Areas north of Everglades National Park. The thin ribbons of asphalt and water that separated neighboring jurisdictions were proving to be of little resistance to the invasion. Rather, they acted as superhighways for dispersal. As scientists continued to offer differing opinions about the ability of the invasive snakes to venture elsewhere, many were already on the move. Burmese pythons were quickly becoming everyone's problem.

———

As wild Burmese pythons began to make their presence known across the broader landscape, novel ideas for control began to circulate. Hoping to tap into the voracious appetites of hungry Americans, some suggested circulating recipes and cultivating a culinary taste for pythons. Elsewhere in the country, specialty shops routinely sell imported python meat for $40–$50 dollars a pound, and market forces would no doubt inspire many to help cull the south Florida population for fun and profit. "Invasivore" diets were increasingly encouraged as a hopeful means of controlling unwanted species like lionfish, Asian carp, and green mitten crabs. In some cases, entire cookbooks had been authored just to promote the movement.

Others have written about the odd paradox of success that can form from such efforts: if exotic species prove palatable to taste and subsequently yield profits, then there can form a strong disincentive to total eradication and a greater inclination toward maintain sustainable populations just to keep business alive. But fears of such an outcome with Burmese pythons quickly became unfounded. The idea of advertising the gastronomic merits of south Florida's wild pythons fizzled in late 2009 when health concerns surfaced regarding their consumption. Analysis of pythons collected from the Everglades bore "extraordinarily high levels" of mercury. A toxic contaminant in the Everglades ecosystem for decades, mercury accumulates in organisms as it travels up the food chain. In the case of pythons, mercury concentrations were up to five times higher than those of alligators. Though not conclusive, the evidence suggested pythons might be routinely making meals of large, top predators in south Florida. And amazingly, the pythons showed no ill effects from the mercury they were amassing within.

In May of 2009, state wildlife authorities began kicking around a new idea in the name of python control—bounties. During a visit to south Florida by Secretary of the Interior Ken Salazar, agency heads offered the idea as an incentive to entice hunters to help in the eradication effort. Historically, bounties played a prominent role in predator control programs that ultimately proved so successful they have left critically endangered populations of wolves and mountain lions in their wake. Today, the relative scarcity of these species tugs at the heartstrings of most Americans. The notion of paid killings, no matter how noble the cause, carries with it both emotional baggage and valid concerns regarding potential abuses by unscrupulous profiteers. After all, if money was to be made in the hunting of large constrictors, it would only make business sense to ensure a steady supply was being stocked for the kill. Might some be willing to intentionally seed a favored spot in the marsh with inexpensive juveniles in the hopes of returning to make a windfall or snare a giant trophy? In Miami-Dade County the notion wasn't a far stretch of the imagination. During the 90s security guards routinely encountered a rogue hunter in camouflage trespassing in a popular park under the cloak of night. In the darkness he would hunt hogs with a large crossbow—hogs, it was suspected, that he himself had been throwing over the perimeter wall as piglets for years.

Perhaps of greater concern, though, was the potential for mistaken identity. Depending on one's perspective, Florida is either blessed or cursed with a multitude of snakes. A total of 28 different species and subspecies exist within the footprint of the southern Everglades, and roughly 80 varieties can be found statewide. Native serpents vary greatly in size, occupy nearly all available habitats, and display every color of the spectrum across a wide variety of patterns. Animal trappers who routinely respond to panicked calls from homeowners have endless tales of innocuous snakes being misidentified as venomous species, or the length of four-foot-long individuals being reported in the double digits. Though exotic in origin, Burmese pythons could prove difficult to differentiate from other native snakes to less familiar eyes. In other parts of the world, errors in identification at the hands of well-meaning individuals have been shown to present potentially lethal impacts to some native populations. The same loss and fragmentation of natural habitats that has impacted the survival of panthers and wading birds is also blamed for the anecdotal disappearance of once-common native species. Now these same species—some already so rare they are protected under federal and state law—might potentially become casualties of well-intentioned friendly fire.

Establishing any open season on pythons—fueled by bounties or otherwise—could also bring unintended consequences beyond mere management. In the American culture of omnipresent food options, hunting has become a largely recreational pursuit—an honored tradition shared among family and friends over generations. For many, time spent in the wilderness is also time spent sharing experiences and valuable life lessons that help forge enduring relationships and memories. And often, these memories are inextricably linked to the targets of their pursuit—their first hog, their biggest buck, or, in Florida, their most ornery alligator. Adding pythons to the list of potential game, therefore, had the potential to promote them to something far greater than a mere public nuisance. Over generations, the act of hunting pythons could easily entrench itself as a staple of the hunting community—a rite of passage as integral to the Gladesmen culture as gigging a bucket of pig frogs or piloting an aluminum airboat. In decades to come, one might expect to have some individuals lobbying to preserve opportunities to pursue pythons with their children—much in the same manner as they had with their own parents. Such scenarios play out daily around the world as authorities struggle to eradicate nonnative species—feral hogs, *Cannabis*, wild horses, *Eucalyptus*, bluegrass, coffee, etc.—from communities where they have found considerable favor with the recipient culture over time.

By all accounts, the idea of hunts and bounties initially received a lukewarm response—particularly from managers of areas like Everglades National Park, where hunting was simply not an option. But attitudes changed quickly following the death of toddler Shaianna Hare in Oxford, Florida. Though the tragedy occurred hundreds of miles from the Everglades and involved a captive snake—not a feral python—some believed the incident foreshadowed what might one day befall visitors to south Florida's watery preserves. Within days, elected officials at all levels were clamoring for greater control. Florida Senator Bill Nelson and Representative Tom Rooney called on Interior Secretary Salazar to support legislation prohibiting the import and interstate sale of the snakes—even calling for a supervised hunt to be conducted by authorized agents in the national park. Within two weeks, Florida Governor Charlie Crist ordered a sanctioned "python posse," composed of select licensed reptile enthusiasts, to begin hunting for the snakes on four state wildlife management areas to the north and east of Everglades National Park. Rodney Barreto, chairman of the Florida Fish and Wildlife Conservation Commission, paraphrased his boss succinctly, "Governor Crist wants to take action to stop the spread of this snake."

The effort began modestly days later with only three permitted hunters on state-managed lands. On July 17, during a media event arranged by the commission to kick off the program, the team snared a nine-foot python off a boardwalk with amazed journalists in tow. The experimental season was to run for three months, after which the efficacy of the program would be evaluated. As a rule, every snake captured was to be euthanized, though the use of firearms was not permitted. Hunters were also required to submit data on their captures to help scientists paint a clearer picture of the problem. The much-talked-about bounty never materialized, but hunters were free to sell the hide of the animals for profit if they so desired. Still, it was the thrill of the hunt, not the money, that spurred most participants into the marsh.

Ultimately, a total of 15 permitted hunters participated in the three-month experiment—to limited success. Intense media interest—stretching from Florida to Los Angeles—helped illuminate the difficulties hunters experienced in the field. "I've been out three times this week and haven't found anything," groused one to a reporter from the *Naples Daily News*. "I'm just not seeing anything. I'm not even seeing signs of anything." Between July 17 and October 31, hunters managed to capture only 39 Burmese pythons. Given a reported 517 man hours of searching, the effort required over 13 hours of time in the field for every one snake found.

Admittedly, the reactionary hunting season was timed all wrong. In the subtropical Everglades, summer yields high water and scorching temperatures—conditions that guarantee snakes will be active, alert, and afforded ample avenues of escape and cover. Hunting would be most effective during times of cooler temperatures, when lethargic snakes were eager to bask in open, sunlit areas. Thus, a second month-long season launched the following March with a well-attended course on python-wrangling offered by state officials. Amidst a flurry of cameras and reporters, several dozen participants learned how to catch and dispatch pythons in the field and were even introduced to commercial processors who would pay cash for their quarry by the foot. At the event, Bishop Wright Jr., president of the Florida Airboat Association, declared on behalf of his fellow sportsmen, "We feel we have the knowledge, responsibility, and technical ability to take care of this problem. We are the best tool in the toolbox in this situation." In the following weeks cameras and notepads would record many similar statements of misplaced confidence. In the field, hunters were ending their days empty-handed—save for stories involving reporters, Africanized bees, and lost wedding rings reminiscent of the Keystone Kops. Though it wasn't clear how many hunters took advantage of the special season, nor how long they had spent searching, the number of pythons captured was certain: zero.

The relatively low catch rate during the organized hunts surprised virtually no one—particularly the biologists who had spent time tracking radio tagged pythons in the Everglades backcountry. When not exposed along roads or levees, Burmese pythons could meld so seamlessly with the landscape that they were often invisible to the naked eye—even when guided by the audible beeps of a planted transmitter. Calls to allow similar hunts in Everglades National Park eventually waned as the limited effect such efforts would have across 1.3 million acres of unimproved wilderness seemed increasingly obvious.

Burmese pythons did not seem destined to join the ranks of Gambian giant pouched rats, red-bellied piranhas, and black-tailed jackrabbits. Eradication seemed improbable, and the lack of any effective management tools seemed to guarantee the ongoing proliferation of the species. Hunting, so often the control strategy of choice for unwanted wildlife, was proving to be of little to no use in the vast expanse of the Everglades beyond well-traveled routes. Worse yet, the hunts themselves were slowly entrenching the species deeper into the ever-shifting realm of normalcy in south Florida.

The organized hunts to ferret pythons out of the Everglades were making headlines across the nation and, in some cases, were again providing readers and viewers with their first exposure to south Florida. As the story unfolded in the pages of *New Yorker*, and as featured content on Animal Planet and the Discovery Channel, more and more park visitors—who in times past would have likely asked about the possibility of seeing alligators and panthers—were now inquiring about where to spot pythons. In previous years, long-standing curiosity about well-publicized disasters in the Everglades was common—the impact of Hurricane Andrew's wrath, for instance, or the location of the 1996 crash site that swallowed a ValuJet airliner and all 110 people onboard when it mysteriously plummeted into the Glades shortly after takeoff from Miami International Airport. Instead, visitors now seemed eager to indulge their morbid curiosity about the biological disaster unfolding in the marsh. And their curiosity would only grow greater when, late in 2009, stories began to circulate about the possible existence of a second species of python taking root in the Everglades.

8
The Big Chill

On a cool, clear day in mid-November of 2009, I packed a pair of gloves, some water, and my GPS for a new adventure. During the previous month, it was made public that several researchers had been monitoring a disturbing trend—regular recoveries of a second species of python from an area in western Miami-Dade County. Since 2002, six northern African pythons (*Python sebae*) had been recovered from a roughly 5,000-acre area referred to as the Bird Drive Basin, a checkerboard of private, county, state, and tribal-owned lands harboring a dense mix of melaleuca forest, rock pits, canals, subdivisions, and remnant Everglades. Recorded captures included several large adults and two small juveniles—what USGS researchers would later call "strong evidence of a reproductive population." During this chilly autumn day, I was to join a motley crew of roughly 30 professional biologists, university professors, graduate students, park volunteers, and amateur reptile enthusiasts on an ad-hoc mission to survey a part of the area in question.

Following a short, early-morning briefing along a lonely road in a sparsely populated suburb of Miami, our group broke apart into smaller teams and ventured due east with the amber light of the rising sun on our backs. We began exploring the margins of a densely overgrown parcel of land known formally as Tree Island Park and Preserve. The length of its borders was marred by the ills that plague all natural areas unlucky enough to interface with the urban world of south Florida. We searched beneath piles of illegally discarded terracotta pots, roofing tiles, broken televisions, furniture, beer bottles, and spent shotgun shells. Surveyors prodded bags and crates containing fruits, clothing, and the putrid remains of chickens—sacrificial offerings made by superstitious practitioners of Santería. Within the protected confines of some dense brush, homeless squatters occupied improvised camps well away from the prying eyes of police. And the abandoned remains of a large horse nearby could only lead us to wonder whether it had recently fallen victim to one of the many illegal slaughterhouses purportedly operating in southeast Florida.

Moving farther from the convenience of nearby roads, the landscape grew less obviously tainted. Among the thick growth of fat-leaved cattails and sharp-edged sawgrass, encounters with plastic and glass became far less frequent. Still, discerning eyes could observe another, more subtle form of pollution causing a blight on the horizon. Dense stands of melaleuca, Brazilian pepper, and ardisia punctuated the otherwise open terrain—all invasive species that run rampant in south Florida and had fared particularly well in this neglected corner of the landscape. The search teams were little deterred by the impenetrable walls of vegetation they occasionally encountered, having come well-prepared for the challenge. My cohorts were fortified with all manner of gear: machetes, snake chaps, walking sticks, snake hooks, empty laundry bags, cameras, and GPS receivers. For the next four hours, our group moved laboriously under rising temperatures, stopping occasionally to peruse, prod, and overturn anything that to us looked even remotely hospitable for snakes. All the while, our collective gaze remained focused downward in a bid to ferret out the newest invader thought to be further polluting this patch of remnant Everglades.

Around noon, all the groups reconvened at the original briefing location for lunch. Hours of searching the area had turned up no pythons. In fact, despite dozens of pairs of scouring eyes, few animals had been encountered at all. No visible evidence suggested pythons were to be found in the area but—if there were any—it was equally dubious whether they could have ever been detected among the luxuriant growth. The results of the morning lured most participants to call it quits for the day. A few of the more committed individuals returned later in the afternoon for a second look. But by sunset, over 100 hours of search time had yielded only a handful of encounters with some of the area's more common reptiles: three ringneck snakes, two racers, and a sole box turtle.

Though searchers were unable to locate a single northern African python that day, few doubted they were present. Encounters with the species continued to get reported and the evidence continued to mount. Perhaps the coming year would bring another opportunity to learn more.

Two years after it had originally been released, the alarming climate suitability study put forth by the United States Geological Survey (USGS) was incorporated into a larger risk assessment developed by the agency at the request of the National Park Service and the United States Fish and Wildlife Service. The report was commissioned as a means of exploring the potential risks to both ecology and human safety associated with the invasion potential of nine large nonnative constrictor species in the United States. Once complete, the published assessment rated all species reviewed (Burmese pythons, reticulated pythons, northern and southern African pythons, boa constrictors, and four species of anaconda) to present either a moderate or high risk of invasion.

Immediately upon its release, the report drew harsh criticism—especially when the work was referenced in subsequent hearings pertaining to proposed regulations on the trade in large snakes. In November of 2009, a consortium of notable veterinarians and herpetologists sharply denounced the study. "As scientists whose careers are focused around publishing in peer-reviewed journals and providing expert reviews of papers submitted to these journals," they wrote, "we feel it is a misrepresentation to call the USGS document 'scientific.' " The group vehemently opposed the use of the study as a basis for legislation or policy, charging that the work was authored to advance predetermined objectives. The signatories advised subjecting the study to the rigors of an external peer-review.

Only two months later, a second group—comprised of distinguished academics from both universities and nongovernmental organizations—drafted a rebuttal letter to members of Congress expressing their unreserved confidence in the USGS report. In their correspondence they underscored the extensive reviews conducted by institutions and agencies outside the USGS and characterized the science behind the study as "reasonable and appropriate." Furthermore, they noted, the study was commissioned only as a guide for future management efforts and not as a vehicle for advancing regulation. As sharply as they had originally lambasted the assessment, the critics themselves were now being criticized for making "unsubstantiated allegations" of impropriety.

Rhetoric from both sides spilled over into the largely unmediated realm of the blogosphere. Responding to an online post on the National Geographic website, the USGS Associate Director for Biology publicly defended the risk assessment and dismissed any claims of bias. The post drew a slew of derisive replies, mostly from anonymous commentators. David and Tracy Barker—the husband-and-wife python breeders from Texas—weighed in on the matter. In a 1,600-word letter to National Geographic, the pair charged the USGS with a lack of impartiality on the topic, given their "decades of experience getting funding for injurious snake research." But while the debate on theory was growing ever more heated, temperatures in south Florida were taking a nose dive.

A historic blast of Arctic air brought south Florida's normally mild temperatures to record lows in early 2010. The stationary front that settled over the Everglades in early January succeeded in bringing the coldest 12-day period on record for the area since at least 1940. The two-week spell brought unfamiliar rituals to south Florida, where people turned on heaters, donned scarves and mittens, and scraped thin layers of ice off frosted windshields.

Climatologically, south Florida is a subtropical region where periodic blasts of cold air are a somewhat regular and necessary occurrence. Occasional, short-lived frosts are said to be one of several environmental factors that help shape the Everglades and maintain the area's diversity—largely by pruning back the growth of vegetation and limiting the reach of tropical organisms from the south. And with regard to nonnative species, they also provide a measure of control against foreign biota not adapted to such extreme drops in temperature.

The January 2010 event was notable, however, for its longevity. For nearly two straight weeks, temperatures averaged around 50 degrees and dropped perilously close to freezing most nights. By the time a warming trend arrived in mid-January, large areas of mangrove forest, hardwood hammock, and pineland had already been converted to frost-bitten stands of gnarled, colorless vegetation. Along the coast, marinas, bays, and waterways were left choked with the remains of tens of thousands of dead or dying mullet, ladyfish, hogfish, bonefish, tarpon, grouper, grunts, and sharks. One researcher alone claimed to have counted about 90,000 deceased snook floating in the waters of Florida Bay during the aftermath. Around the state, rescuers attempted to relocate hundreds of cold-stunned sea turtles pulled from frigid, near-shore waters to rehabilitation facilities nearby. And aerial surveys revealed the grisly toll cold exposure had taken upon some of south Florida's most iconic wildlife: over 240 manatees were discovered floating belly-up and dozens of crocodiles were found dead.

But in the aftermath of the big chill, biologists also made note of some positive impacts. Two problematic invasive plants, Old World climbing fern and Brazilian pepper, had taken a hard hit—particularly along the coast. Within the freshwater marsh, a wide array of nonnative fish—Mayan cichlids, African jewelfish, walking catfish, spotted tilapia, and Asian swamp eels—had succumbed to the cold. And in residential neighborhoods, parks, and on golf courses across south Florida, green iguanas, Cuban knight anoles, and brown basilisks were dropping out of trees like so many ripened mangos, providing a valuable opportunity to capture the wily pests with ease.

In the wake of south Florida's rare brush with winter, reports began to circulate about the discovery of several sick and deceased pythons. The bitter temperatures had unexpectedly provided a living laboratory with which to referee the ongoing debate about how resistant Burmese pythons were to cold, and how far the species might be expected to spread northward. At the time the mercury fell below freezing, there were three ongoing, concurrent studies that left Burmese pythons exposed to the elements.

Within the walls of their outdoor enclosure in South Carolina, the ten pythons housed at the Savannah River Ecology Lab had already endured some unusually cold weather for the region. Following an appreciable dip in temperatures beginning that October, the snakes began to abandon their watery haunts of summer to instead bask in the sun or occasionally retreat to subterranean refuge nearby. And as autumn turned to winter, and temperatures continued to fall precipitously, the snakes began to fulfill their intended purpose. The pythons increasingly struggled to maintain their body temperatures, all the while providing valuable data to a gaggle of observant researchers. But as the blast of Arctic air made its way down the Florida peninsula, additional pythons far from the flood plains of the Carolina coast were also becoming barometers of cold tolerance.

In early 2008, nine wild-caught pythons from the Everglades had been relocated to an outdoor laboratory in Gainesville. These pythons—three females and six males—were all being kept at the Florida Field Station of the U.S. Department of Agriculture, where they served as test subjects for experimental trials involving reproduction, trapping strategies, and the use of chemical cues. Each of the large aluminum pens in which the snakes were kept featured an array of reptilian creature comforts—a plastic water pool, basking surfaces, and an insulated wooden hide box complete with heating pad. The furnishings proved sufficient enough that all nine adults survived outdoors through one full winter the previous year. But the cold snap of 2010 raised the stakes significantly and provided an unexpected opportunity to test the physiological limits of pythons in central Florida.

Closer to the Tropic of Cancer, the radio tracking of wild pythons in the Florida Everglades was ongoing. At the time the bitter cold of 2010 hit south Florida, ten snakes were being tracked with telemetry equipment. The snakes were not only implanted with a pair of transmitters but, like their brethren in South Carolina, were also outfitted with a tiny instrument to record body temperature. During the prolonged freeze, researchers located each subject one to two times per week until they could confirm either death or survival. And over the course of that month, researchers simultaneously surveyed the roads, trails, ponds, levees, canals, and upland ecosystems in Everglades National

Park looking for additional pythons in distress.

By mid-January south Florida's famously mild climate rebounded, much to the relief of both residents and tourists. And it appeared the local community could celebrate more than just the return of beach weather—evidence suggested that the cold had taken a heavy toll on the population of pythons in the Everglades. Of the ten tagged pythons recovered in Everglades National Park, only one managed to survive. What's more, it didn't seem pythons would be making too many inroads to the north. Of the nine snakes kept outdoors in Gainesville, only two survived without showing ill effects. And at the Savannah River Ecology Lab, where it was hoped ten Burmese pythons could provide insight into their potential for cold tolerance, the outcome was unanimous: all ten perished within the month culminating in the January cold snap.

In subsequent reports published in the journal *Biological Invasions*, observers noted a behavioral similarity among pythons across all three studies. That is, in the face of freezing temperatures, most snakes eventually made really bad choices. Rather than seeking shelter in the warmth of underground refugia or hunkering down in the confines of a heated hide box, most attempted to either bask during freezing temperatures or regulate their body temperature by submerging themselves in shallow water. Yielding lethal results, these "maladaptive" behaviors have been interpreted by some as clear evidence of an innate climatological bias in Burmese pythons that leaves them ill-prepared to contend with periods of intense cold.

The preliminary numbers inspired some to jump at the chance to eagerly spread the good news. "Anecdotally, we might have lost maybe half of the pythons out there to the cold," reported one spokesman for the Florida Fish and Wildlife Conservation Commission. Those with interests in safeguarding the private ownership of large constrictors trumpeted the mortality numbers as a means to help alleviate fears of a northern migration. In speaking to reporters, Andrew Wyatt, president of the United States Association of Reptile Keepers, said of the data, "It reinforces the idea that the pythons can't exist more than a short period of time north of Lake Okeechobee." In cyberspace, message boards dedicated to reptile husbandry resonated with lengthy posts that—while lamenting the heavy death toll—insisted the tragic outcome had invalidated the need for additional regulation. Others, however, cautioned about the dangers of extrapolating data gained from a very small sample size across a potential population in the tens of thousands. What's more, the nuances of each study demanded more meticulous scrutiny.

After all, it was the outliers from each study that were of particular concern. In Gainesville, two snakes survived, showing no visible sign of distress. In Everglades National Park, one of the telemetered snakes managed to make a wise move—seeking shelter in the humid warmth of a forested subtropical hammock. And researchers surveying the park for non-tagged snakes eventually found a total of 104—of which 60% had survived the bitter blast at the time of discovery. Most of the wild snakes that survived the ordeal were noted to be associated in some way with artificial habitats, making use of the myriad levees, canals, and roads, the likes of which crisscross the whole of Florida. But otherwise, there was little in the way of biology to differentiate survivors from those that succumbed—no apparent correlation among sex, length, age, or weight. It was unclear why, or how, those that lived to see February managed to do so, but understanding the nature of this feat was important. If there was some physiological advantage that allowed some to survive, it might be tied to a particular trait which, in the course of reproduction, could be handed down to progeny. If the ability to survive is somehow linked to heritable genetic code, the ecological effect of the 2010 cold weather event changes considerably. Rather than a friendly assist from Mother Nature in culling an unwanted invader, south Florida may have experienced a potent demonstration of natural selection, wherein future generations would be that much more cold tolerant.

There remained much fodder for discussion. Owing to the relative lack of genetic diversity previously discovered within the south Florida population of Burmese pythons, some questioned the potential for adaptation over time. Some wondered whether the extraordinary temperatures experienced in Gainesville and South Carolina fairly represented year-to-year realities, or if the studies might have garnered different results by offering deeper refugia.

Others noted that, in some snakes, appropriate regulatory behaviors are imprinted very early in life—leaving adults with an inability to contend with climatic conditions beyond those experienced in their youth. Perhaps the fates of the adult pythons born of the Everglades and shipped to points farther north had already been predetermined.

During an interview the following fall, researcher Skip Snow responded to all the uncertainty and prognostication saying, "The snakes are going to tell us. They're clearly here, and they're breeding, and they're established and they're going to tell us over the years and over the decades just what they can put up with and how far they can go." What else might go along with them, however, remained unanswered.

Bob Hill has spent almost 40 years working as a maintenance employee for the South Florida Water Management District. During his tenure, he has driven the miles of levees that frame and bisect Miami-Dade County thousands of times. In more recent years, however, he has earned two big distinctions within his profession. He is currently the only employee in his agency authorized to carry a firearm, which, in this case, is a Winchester semi-automatic shotgun. He carries the weapon in service to his other distinction—south Florida's most successful python hunter. Hill has single-handedly killed more of the unwanted snakes than any other individual. Since 2004, he estimates he has captured over 300 pythons on the job.

Hill is distinctive in his appearance—white whiskers frame his rounded face and a blue denim shirt, matching pants, and well-worn work boots often cover the length of his stout frame. A boldly imprinted South Florida Water Management District cap often provides the only indication that Hill is on the job.

In January 2010, in the thick of the bitter cold snap, I joined Hill for a second organized survey for northern African pythons. Sightings had continued to accrue from the westernmost outskirts of Miami—a wildlife officer captured an irritable adult in that area only one month before. Over the next three days, government biologists would work alongside reptile enthusiasts, law enforcement officers, and field technicians to survey choice locations over approximately 5,000 acres of land. In the months of planning and preparation that preceded the event, organizers could have never guessed Mother Nature would provide such ideal conditions for the effort.

Early on the morning of January 12, 40 participants gathered not too far from the speeding traffic along Tamiami Trail for an initial briefing on the first day. Bundled in jackets and sipping hot coffee, participants stood at attention, snake hooks in hand, as organizers provided details about the survey area, counseled on how to gather the requisite data, and reminded everyone about the importance of providing for their personal safety. Then a live—and incredibly lethargic—Burmese python was pulled from a canvas bag and used for an impromptu lesson on how to properly capture large snakes without getting bitten. With every breath, condensation emanated from the mouths and nostrils of the participants, who were now fidgeting—from the cold, perhaps, but more likely in a subconscious display of eager anticipation.

As the morning pep talk concluded, the group quickly broke apart into smaller teams—each equipped with a camera, a GPS, snake bags, and a cell phone programmed with the number to call should a python be found. With that, each group departed from the staging area to begin scouring the targeted sawgrass marshes, melaleuca forests, rock pits, brush piles, canals, levees, and residential subdivisions nearby. To do so, some drove vehicles along levees and roads bordering the survey area, while others explored on foot, paddled canoes, or rode ATVs.

After only two hours of searching in the field, an excited phone call was received back at the briefing area. A twelve-foot female northern African python had just been discovered in a brush pile nearby. Thanks to a cold-induced torpor, the 64-pound snake was easily captured and loaded into a hard plastic container, which was then strapped to an all-terrain vehicle for a ride back to the staging area. Upon arrival, the animal was photographed, measured, weighed, and prepared for transport to a nearby laboratory. Officials marveled at the size and apparent health of the python. Had she been living alone in the marsh, or were others to be found nearby? The answer came only a few hours later when, not far from where she had been captured, a second team called in to report that they had found a monster.

One survey team had been following the overgrown banks of a remote green-water canal, when the ground beneath them suddenly moved. From the tangle of underbrush, the team unearthed a whopping 14-and-a-half-foot male—still the largest northern African python ever recovered from the wild in Florida and likely the largest male ever recorded for the species (Figure 16). At nearly 140 pounds, and measuring two feet around, the team was unable to fit the snake into even their largest collection bag. Instead, the group opted to secure the head—cinching a bright yellow mesh bag as tight around as possible. The group then tackled a bigger question—how to transport this hefty behemoth from their remote location to the distant staging area. The decision was made to launch a canoe in the nearby canal and attempt to motor the animal back to the nearest roadway. Five people labored strenuously to lower the weighty python over the steep banks of the canal and into the red-hulled vessel waiting below—the bulk of its mass still unsecured save for its head. Had such a feat been attempted in the heat of a typical summer day, the snake would have easily flailed free. But in the throes of such bitter cold, the serpent could scarcely stage a revolt against the actions of his captors. In placid repose at the bottom of the vessel, the snake would enjoy a short cruise to the busy highway nearby. As the canoe got under way, I was asked to meet it upon arrival and help load the frigid beast into a waiting government vehicle for transport.

I watched with saucerlike eyes as the vessel motored toward me with its reptilian cargo. Thick coils of scaled flesh left little indication where the serpent began or ended, save for an occasional glimpse of yellow from the bag around the animal's head. Under considerable strain, three of us maneuvered the unwieldy snake from the canoe, onto the canal bank, and finally into the back of a large, white Ford Bronco. Throughout the process, the python remained motionless. Still, as three o'clock loomed near and the afternoon sun brought welcome warmth, some worried about how our captive's temperament might begin to change. Given the lack of means to safely contain the Goliath, and knowing that the animal was not long for this world, the decision was made to dispatch the animal prior to transporting it to the laboratory. In a bid for both good taste and public safety, it was decided the deed would be done well away from the busy highway. With the python unconstrained and nearly immobile in the back of our SUV, we drove down a dusty white gravel road atop a nearby levee until we were well out of view. There waiting for us, dressed in his trademark blue and standing aside his white South Florida Water Management District truck, was Bob Hill.

Hill lent formidable strength as my teammate and I struggled to extricate the unwieldy snake from the back of our vehicle. Stretched to its full length, the python nearly spanned the entire road bed. After taking several pictures, we positioned the snake closer to the edge of the levee in preparation. I remember crouching low and holding the drawstring of the mesh bag around the animal's head—attempting to exert some control and awaiting orders to perhaps remove the bag or clear the way. Behind me, I heard Hill preparing his shotgun for duty. Then, unexpectedly, he unceremoniously instructed me to hold still. Hill fired point blank over my right shoulder—the blast leaving my ears ringing and the proximity of the barrel leaving me slightly unnerved. It was a clean shot—testimony to Hill's skilled marksmanship. Within seconds, still holding the drawstring in my hand, the bright yellow mesh bag moistened to a dark shade of crimson.

Copious amounts of blood leeched from the gaping wound behind the animal's nearly severed head. The formerly motionless body now writhed, prodded into motion by the force of the trauma inflicted upon it. I cannot clearly recollect how long we waited—allowing motion and bleeding to slow—before loading the pitiful creature back into our car. I only remember feeling utter sadness at witnessing, and being party to, the violent execution of such a magnificent animal. Before returning to ride shotgun back to the staging area, I bent down and picked up the bright-red shotgun shell cast off from the lethal discharge.

Now early evening, all the crews had reported back in from the field and were awaiting our return to the staging area to lay eyes on the record-setting python. The carcass was unfurled once more for photographs before being packed up for transport. As the sun waned, and temperatures began to plunge once more, the search teams slowly began to disband. As they did, many probably left wondering what else they might find when they regrouped the next morning.

Over the full three-day survey period, the teams would ultimately collect a total of five adult northern African pythons—nearly doubling the number ever recovered from the area. The captures provided additional evidence pointing to an established, reproducing population on the easternmost edge of the remnant Everglades. The survey teams also encountered an array of exotic bycatch—Cuban knight anoles, southern brown basilisks, a Burmese python, and a red-tailed boa. Surprisingly, one of the largest snakes caught during those three days wasn't found anywhere near the survey area. Roughly 250 miles to the north, game officers in central Florida stumbled upon a 12-foot green anaconda near death in a pond in Kissimmee. The same bitter cold snap that had betrayed south Florida's pythons was unexpectedly turning up new monsters elsewhere.

9
Legislation-Come-Lately

In 2005, I purchased a print from a volunteer photographer at Everglades National Park that now graces the guest bathroom in my home. During his tenure, the seasonal shutter bug was able to capture several striking images of a Burmese python resting upon the furrowed branches of what appears to be a young buttonwood tree. Though pythons are occasionally known to climb trees aided by a strong prehensile tail, photographers are rarely lucky enough to catch them in the act. The photo was almost certainly staged.

As if the sight of a constrictor dangling precariously from a limb was not compelling enough, the photographer manipulated the image using a well-worn Photoshop technique. The photograph now nestled within a solid wood frame on my bathroom wall is completely black and white, save for the strikingly conspicuous figure of a full-color, golden-brown serpent. It is an effective conversation starter, as it always generates comments from guests. I've often considered adding some text beneath it to read, "If you sprinkle when you tinkle, wipe the seat or you I'll eat."

If only the colors of the real world could be manipulated and muted so easily. In the vibrant, multi-hued environs of the Everglades, the Burmese python remains cryptic and nearly invisible to human perception. If one could only Photoshop the sawgrass marsh—render the landscape in simple black and white—removing pythons would be a simpler task. Like prospectors sifting through drab mounds of dirt in search of precious metals, we could train our eyes to seek out brief flashes of color. New technologies and methods currently in development might one day provide us with better means of detection but, for now, pythons continue to enjoy a covert existence in the marsh.

Even if such capabilities were available to us in the present, previous experience with the management of other nonnative species provides little hope for optimism. In Florida, a land plagued by hordes of plants and animals that thrive in plain sight, only a few victories can be claimed in the quest to eradicate invasive species. The rare confluence of favorable circumstances that facilitated these success stories has certainly proven the exception and not the norm, as literally hundreds of other species continue to persist by exploiting every favorable niche in the ecosystem.

Yet what if the dogged determination to eradicate the Burmese python were aided in every possible way by both technology and Mother Nature? What if the campaign to stop the spread of this insidious serpent were to ultimately prove successful? The effort would most certainly take decades at least, but wouldn't it ultimately be worthwhile? Couldn't we finally rest at ease knowing that pythons would no longer interrupt the balance of life as we know it? Well, not really.

Every passing year, new species are added to the ever-burgeoning list of invaders now calling south Florida home. New exotic organisms are introduced far faster than others can be effectively controlled. And where there are no prohibitions regulating the importation, sale, or ownership of known problematic species, new introductions constantly threaten to extend the range of established species beyond their original area of infestation, or permit colonization over entirely new areas elsewhere.

The struggle against invasive organisms is a battle with almost endless fronts that threatens to divide and conquer resource managers operating within the constraints of limited funding, personnel, and jurisdictions. Indeed, trying to keep up with ceaseless new invasions is akin to keeping pace with a treadmill that runs a little quicker with every passing year. The frequency with which new invasions occur mandates that the management of invasive species be somehow prioritized. Typically, those organisms that either pose the greatest threat to the ecosystem or present the greatest chance for control are afforded the lion's share of attention. At the other end of the spectrum are those invaders that are altogether ignored either because their perceived impacts are negligible or because they are already so widespread and numerous that any efforts at management are regarded as futile. Meanwhile, hundreds of foreign plants and animals continue to reside in a state of limbo between these two extremes—paid mind only whenever the rare opportunity presents itself. And in some instances, management efforts are precipitated largely by the relative fame—or infamy—of the invading species.

Halting the ongoing introduction of new species is widely regarded as a necessary prerequisite for gaining ground in the struggle to control invasive biota. Across the country, countless public education and awareness campaigns have been developed in a bid to help stop the spread of new species *before* they become bitter enemies entrenched on the landscape. Thwarting new invasions before they begin is understood to be the most effective, and least expensive, strategy against invasive species. And as control costs spiral and management budgets stretch ever thinner, hindsight reaffirms the truth that an ounce of prevention is worth a pound of cure.

Public education efforts, however, have clearly proven themselves limited in their efficacy. Despite years of concerted effort, the introduction of new species continues unabated. Consequently, well-crafted legislation and regulatory mechanisms seem increasingly necessary as a powerful tool for curtailing the threat of invasion presented by the ceaseless waves of imported organisms that land on our shores annually. In the case of the Gambian giant pouched rat, for example, the speedy enactment of an emergency prohibition on their sale bought enough precious time for managers to eradicate the nascent population without the threat of further recruitment.

Such examples highlight how the regulation of trade in certain species potentially has the opportunity to play an important role in protecting ecosystems. And yet, broadly enacting such safeguards has historically proven difficult. The United States remains the single largest importer of live animals in the world. In the few short years since I purchased my python photograph, well over one *billion* new animals have legally made their way past our borders. In the absence of a comprehensive strategy to prevent the escape, establishment, and spread of new species, I wonder what future portraits are destined to grace my bathroom walls.

In 2007, 32-year-old Tania Dumstry-Soos was mauled to death by a Siberian tiger at a privately owned roadside zoo in British Columbia. Gangus, one of three tigers housed at the facility, was able to reach just far enough beyond its locked, 12-foot-by-12-foot chain link cage to grab at the woman's legs and begin pawing furiously. The beast savagely clawed the mother of two as several children looked on in horror—including one of her own. As she lay dying beside the animal's enclosure awaiting the arrival of responders, Dumstry-Soos mustered just enough strength to utter her final goodbyes over the phone to the owner of the facility, her new fiancé, Kim Carlton.

In February of 2009, a 200-pound chimp named Travis violently attacked Charla Nash in the Stamford, Connecticut, home of friend and owner Sandra Herold. The horrific assault, which ended only after responding officers shot the animal, left Nash near death with extensive injuries to her face and hands. The severity of the attack garnered worldwide attention, and the brutality of the ordeal left Nash so disfigured that Stamford Hospital provided psychological counseling for the staff members who provided her care. Nash would spend the next two years suffering through a string of surgeries—including full face and hand transplants—in an attempt to repair the damage.

Only one year later, another animal attack made headlines. On February 24, 2010, trainer Dawn Bransheau was dragged to the bottom of a deepwater tank by Tilikum, a 12,000-pound male orca, following a show at the Sea World amusement park in Orlando, Florida. Reporters were quick to uncover that, long before pulling the 40-year-old trainer under by her long ponytail, the same whale had previously been implicated in the drowning deaths of two other people. Ironically, it seemed it was not so much the animals at SeaWorld that had been trained, but the public—those genuinely surprised that a six-ton killer whale might occasionally revolt against years of training that endeavored to suppress its natural instincts in the name of family-friendly entertainment.

Several months later, the dangers of believing that wild animals can become fully domesticated were illustrated yet again as tragedy unfolded in Ohio. World Animal Studios, located southwest of Cleveland, was a private menagerie containing dozens of wolves, tigers, and other exotic beasts collected by owner Sam Mazzola. In August of 2010, one of Mazzola's hefty pets—a 400-pound black bear— attacked and killed 24-year-old Brent Kandra after it had been purposely freed from its enclosure for a feeding. Long before the incident, Mazzola had raised the ire of animal rights activists not only for amassing multiple convictions of illegally selling and transporting wildlife, but also for his history of famously offering paying customers the opportunity to enter a ring and wrestle his bears.

In aggregate, such stories provided powerful testimony to the relative ease with which exotic animals can be acquired, the often lax manner in which their ownership is regulated, the varying degrees of competency that inform their care, the frequency with which errors can be made, and the sheer randomness that often accompanies attempted relationships with large, dangerous, and predatory wildlife. Though well-publicized when they occur, human fatalities from captive wildlife are still relatively rare. What is more surprising, however, are the number of grave errors regularly committed—and seldom reported—that carry the potential for similar harm to animals, keepers, communities, and the environment.

Media outlets have reported on a steady stream of incidents involving nonlethal attacks, animals rescued from deplorable conditions and negligent owners, and the occasional escape of potentially dangerous species—like the 15 patas monkeys that successfully absconded from a Tampa wild animal park by swimming across their 60-foot-wide retaining moat in April of 2008. Or the wayward tiger rattlesnake that quietly crawled away from Zoo Atlanta in August of 2010 only to be found days later by two-year-old Pierce Mower as it lounged on the porch of his home nearby. And who can forget the Twitter-fueled saga of Mia, the Egyptian cobra that went missing for a week from the confines of her enclosure at the Bronx Zoo in early 2011? While all of these stories ultimately shared happy conclusions, it is not difficult to imagine the myriad ways such escapes could have ended.

In 2010, in an effort to draw greater attention to the issue, Born Free USA launched an online searchable database of incidents involving captive wildlife.

To date, the group has identified over 1,500 cases of abuse, escapes, attacks, and fatalities since 1990. Interestingly, the tally includes 75 human deaths, the most commonly reported species are reptiles, and Florida leads the country in the number of captive animal incidents recorded.

The apparent frequency of serious incidents arising from captive wildlife—many of them involving large constrictors—increasingly garnered more and more attention over the years. And yet there seemed no end to new stories of negligence or carelessness by owners of pythons in Florida. In 2011, a 14-foot Burmese python captured loitering near an apartment complex in Tarpon Springs was discovered to belong to Scott Konger, owner of a nearby aquarium from which the snake had escaped two years prior. In 2009, Justin Matthews deliberately planted his 14-foot Burmese python Sweetie in a culvert near a day care center in Bradenton, then staged an elaborate capture for arriving cameras—a ruse for which he would be charged with a third-degree felony and a second-degree misdemeanor several months later. And that same year, Lemuel A. DeJesus was arrested when officers responding to a tip arrived to find an uncaged, 11-foot-long Burmese python roaming his Crestview, Florida, home. Particularly troubling was the realization that several mattresses shared floor space with the snake, upon which investigators found children's clothing. "With a snake that size," commented one officer, "that's just a disaster waiting to happen."

While it has been suggested by some that the advent of the 24-hour news cycle now foments an unfair portrayal of captive animal attacks as commonplace, it is also equally possible to suggest the expanded news cycle simply provides more ample time to expose a troublesome reality. Whatever the reason, gruesome fatalities at the hands of captive wildlife were catching the attention of Americans through headlines blaring from major media outlets across the nation. At the same time, Burmese pythons were gaining considerable notoriety as one of the most daunting threats imaginable to the south Florida environment. And in both cases, the irresponsible ownership of large predatory animals was being excoriated as the cause. Following her death, two-year-old Shaianna Hare became a catalyst that would blur the lines between human safety and environmental well-being and congeal demands to safeguard both through tighter controls on the trade in large constrictors.

In the face of increasing concerns over human safety, costs of exotic animal control, and the threat of ecological harm, many municipalities and states introduced ordinances to regulate the ownership of non-domesticated animals. Some of these efforts successfully banned the ownership of dangerous species from the suburbs of Chicago and the state of Rhode Island. Others, like those in Ravenswood, West Virginia, imposed permit requirements or fees. Still others, like measures passed by the legislature of North Carolina, imposed strict caging and transportation requirements. In 2008, the Texas legislature passed measures requiring the acquisition of a permit to possess or sell green anacondas and any of four large python species. And in Ohio, in response to the tragic mauling of Brent Kandra by one of Sam Mazzola's captive bears, outgoing governor Ted Strickland issued an executive order enacting a temporary restriction on the possession, sale, and transfer of dangerous wildlife in the state.

In response to the brutal death of Tania Dumstry-Soos by Gangus the tiger—and under great influence by the local Society for the Prevention of Cruelty to Animals—authorities in British Columbia enacted sweeping amendments to the regulation of exotic animal ownership under the Wildlife Act. The new Controlled Alien Species Regulation prohibits the breeding and personal ownership of approximately 1,200 exotic animals throughout the province, seeking to both allay public safety concerns and better protect the interests of captive wildlife. The new regulations took effect on April 1, 2010, and violators now face stiff penalties of up to $250,000 in fines and/or two years in jail—not to mention the seizure, relocation, or destruction of their animals at their own expense.

In Florida, state authorities have reason to be cautious when it comes to exotic wildlife. Jutting awkwardly into the Caribbean, the peninsula lies at the confluence of many factors that facilitate their introduction and spread.

In his book *Animal Underworld: Inside America's Black Market for Rare and Endangered Species*, author Alan Green reminds us why this is so:

> Florida has ample space to accommodate its legions of breeders and dealers. It is estimated that Miami is the port of entry for three-quarters of the nation's legal wildlife imports and for most of the illegal shipments. The mild climate makes the state a wintertime dumping ground for northern petting zoos. Theme parks, roadside menageries, and tourist traps lure customers with exotic species. And an amalgam of immigrant cultures has fueled the demand for an unusually diverse array of pets and animals used for less noble purposes.

Perhaps born of the need to respond to this dynamic, Florida is often credited with enacting some of the most stringent laws on wildlife ownership in the nation. In 1967—largely in the interests of ensuring public safety and animal welfare—officials initiated a series of captive wildlife regulations that persists in modified form to this day. Presently, regulation is accomplished through a tiered permit system that is somewhat unique in the United States, in that it not only places blanket restrictions on the possession of all wildlife, but subsequently also lists species for which permission to own may be granted in the form of a state-issued permit.

Generally, listed species are divided among four categories in accordance with the relative risk they pose to people, though how these lists are arranged can be somewhat difficult to understand. For example, *Class I* species (chimpanzees, lions, bears, elephants, Komodo dragons, etc.) have been deemed too dangerous for personal possession and, with few exceptions, are generally restricted to ownership by zoos and research facilities. *Class II* species (smaller monkeys, cats, ruminants, etc.) are those judged to present a more moderate risk to people, and as such may be permitted for personal possession. *Exempt* species—mostly smaller mammals and birds (gerbils, parrots, rabbits, finches, etc.) are those commonly kept as household pets and generally not thought to present a risk of harm. And *Class III* wildlife, though no formal list exists, includes all animals not currently listed as *Class I*, *Class II*, or *Exempt*.

Depending upon the specifics of the animal, permit holders for *Class I* and *Class II* wildlife may be required to pay annual registration fees, demonstrate sufficient past experience in the husbandry of the species, file detailed critical incident action plans, adhere to strict caging requirements that protect both the welfare of the animal and the general public, or provide documented approval from the appropriate municipal or county authority. Similarly, owners of most *Class III* wildlife are required to meet minimum age standards, adhere to animal welfare requirements, and possess a no-cost state permit.

But under the system, the personal ownership of *Exempt* species was largely unregulated, and included in the list of such organisms were all "reptiles and amphibians." That is to say, with the exception of Komodo dragons, crocodilians, and a handful of tortoises, Florida statutes issued blanket permission for the sale and possession of *every* nonvenomous reptile species—including the largest constrictors in the world. Others have written eloquently about how overworked wildlife agencies are often forced to prioritize their efforts to the benefit of agricultural livestock, native game, or endangered species—leaving other taxa largely unregulated. This trend, however, often changes abruptly in the face of crisis. Yet for many years, neither the escalating arrival and establishment of new species documented by periodic herpetological surveys, nor the frequency with which wayward snakes and lizards were being encountered in residential neighborhoods and suburban parks, proved reason enough to enact more stringent rules on ownership.

But as encounters with Burmese pythons in Everglades National Park continued to escalate, many began to question whether more should be done to prevent the litany of intentional and accidental releases that had been occurring throughout the state for decades. After all, invasion biologists were increasingly beginning to find that one of the most important factors dictating successful invasion was "propagule pressure"—what author Alan Burdick has called "the frequency and persistence of introduction." That is, while one might expect most organisms to fail when abruptly thrust into new surroundings, some small percentage will likely succeed if environmental conditions are right. And regardless of how slight a percentage that might be, the greater the number of individuals introduced betters the chances a population might prosper. The idea can be illustrated by imagining a novice gardener who—frustrated that his efforts have not yet born fruit—continues to sow seeds of every variety, upon every inch of soil, at every time of year. Cast enough seeds and, eventually, *something* will take root and flourish.

Florida's canals and lakes—which now harbor a colorful amalgam of foreign fishes and invertebrates that, in some areas, outnumbers the remnant natives—are a testament to the effects of frequent and persistent introduction. Decades of suspected releases from aquaculture and personal aquarists have resulted in the establishment of thriving populations of walking catfish, Asian swamp eels, Mayan cichlids, oscars, bullseye snakeheads, Mozambique tilapia, African jewelfish, brown hoplos, and island apple snails, to name a few. In the 1970s, seeking to stem the tide of illegal releases of potentially damaging species, the State of Florida moved to prohibit the personal possession of certain aquatic organisms—particularly those that had already demonstrated their potential for harm. Listed as *Conditional* species, these organisms could still be possessed for commercial purposes, but were subject to strict oversight. Though some organisms (i.e. freshwater stingrays) were listed as *Conditional* for the purpose of human safety, others (Nile perch, Asian carp, etc.) were now being regulated for their invasive potential on the grounds they might present a hazard to the local ecology.

Beginning in 2008, the State of Florida began to take measures to better regulate the ownership of several large reptile species and help prevent new introductions from occurring. That year, officials categorized the Nile monitor, green anaconda, and four different species of python as *Reptiles of Concern*. Those wishing to personally possess any of these species would now be required to be at least 18 years of age, purchase an annual $100 permit from the state, demonstrate a knowledge and proficiency in working with the species, file a detailed disaster/critical incident plan, and have the snake implanted with a PIT (passive integrated transponder) tag to assist in identification in the event of escape. State officials also began routinely offering regular nonnative wildlife amnesty events, where owners unable to comply with the new regulations (or harboring entirely prohibited species) could surrender healthy animals to a network of qualified adopters without question or fear of prosecution.

The new regulations would ultimately bring mixed success. In the first two years following the rule change, the state issued permits for fewer than 400 *Reptiles of Concern*. And though owners had not been formally tallied in years prior, it was almost certain many thousands more were likely keeping such species within the state. To some critics, the new regulations were not only overdue but, ultimately, proved to be too little too late. After all, three of the six species listed were already present and breeding in the wilds of Florida, while the boa constrictor—already an established invader on the southeast coast—had been entirely ignored. Florida's efforts to prevent the spread of invasive species, like those of other states and federal agencies, have been criticized as being too reactive in nature and not sufficiently preemptive. Still, many welcomed the new rules as a necessary first step, and applauded the foresight of regulating several additional large constrictors that, while not presently on the loose, could be similarly capable of invasion.

But only one year later, the highly publicized death of two-year-old Shaianna Hare would bring renewed calls for even stricter legislation. In 2009, Eleanor Sobel introduced a bill into the Florida Senate that sought to immediately enact an outright ban on the possession of all *Reptiles of Concern*. The bill, whose companion was introduced to the House by Representative Trudi Williams, easily cleared the legislature by early 2010. In anticipation of the pending change, the Florida Fish and Wildlife Conservation Commission acted quickly to organize a year-round amnesty program for *Reptiles of Concern* that would help encourage the responsible surrender of illegal pets and prevent more releases into the wild. And before being signed into law by then-governor Charlie Crist, state officials drafted a provision aimed at protecting reptile business interests by continuing to allow for trade between breeders and out-of-state buyers.

In summer of 2010, Crist formally signed the bill, which reclassified all former *Reptiles of Concern* as *Conditional* species, thereby prohibiting them as personal pets statewide. Provided they had complied with all permit

Figure 1. Author (left) with eight-foot Burmese python surrendered at the Ernest Coe Visitor Center in 2001. Image courtesy of the National Park Service

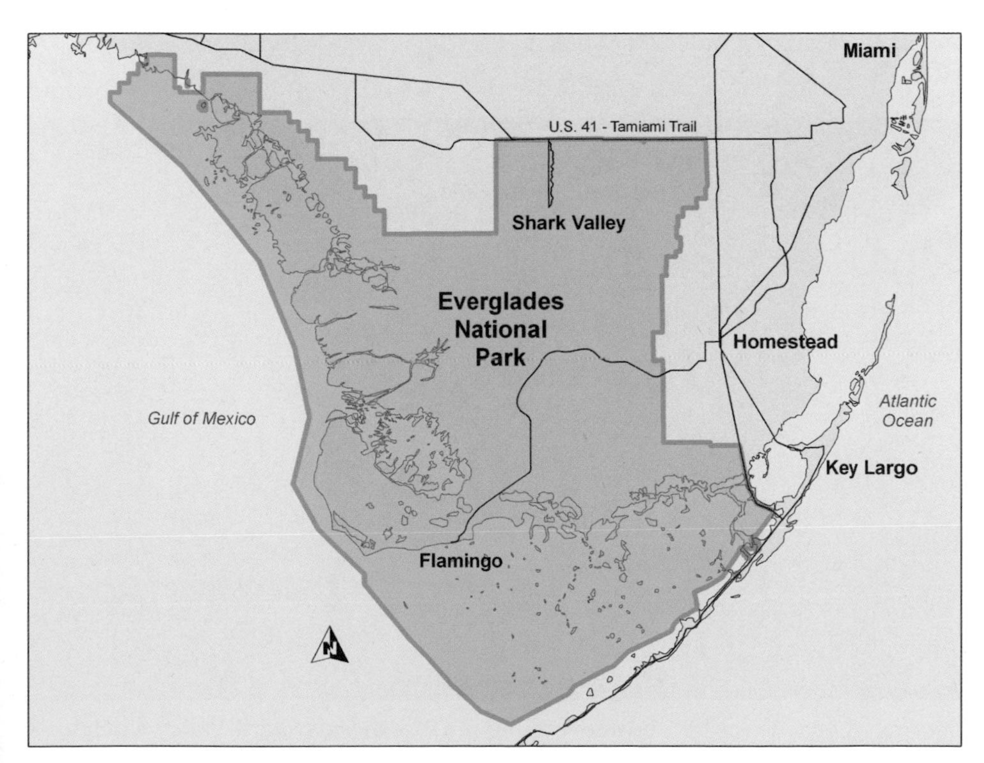

Figure 2. Map of Everglades National Park. Data courtesy of the National Park Service.

Figure 3. Since 2003, visitors to Everglades National Park have occasionally witnessed alligator/python interactions. Image courtesy of the National Park Service.

Figure 4. A grisly scene was discovered in the marshes of Shark Valley when a twelve-foot python unsuccessfully attempted to consume an adult alligator in 2005. Image courtesy of the National Park Service.

Figure 5. Encounters between people and Burmese pythons began escalating in the early 2000s. Image courtesy of the National Park Service

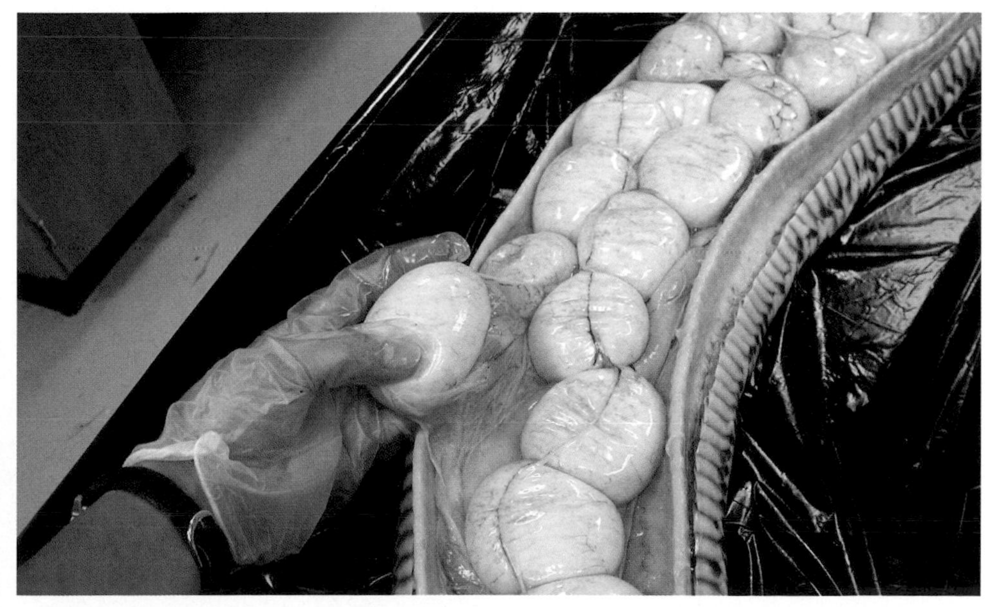

Figure 6. Necropsies reveal the presence of egg-laden females among the pythons recovered from the Everglades. Image courtesy of the National Park Service.

Figure 7. Jim Duquesnel, Joanne Potts, and Clay DeGayner pose next to the Burmese Python found on Key Largo. A subsequent necropsy would reveal the serpent consumed not one, but two, endangered Key Largo Woodrats. Image courtesy of Jessica Demarco.

Figure 8. Burmese pythons are known to subdue and consume a wide variety of prey, like this unfortunate White Ibis. Image courtesy of Tim Taylor.

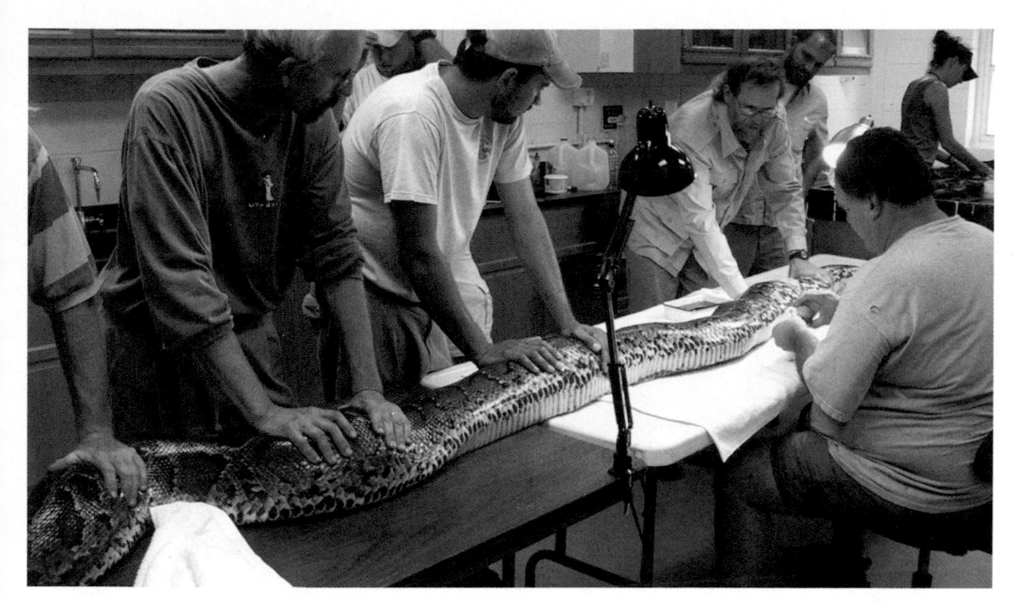

Figure 9. Team implants female python with radio transmitter for tracking. Image courtesy of the National Park Service.

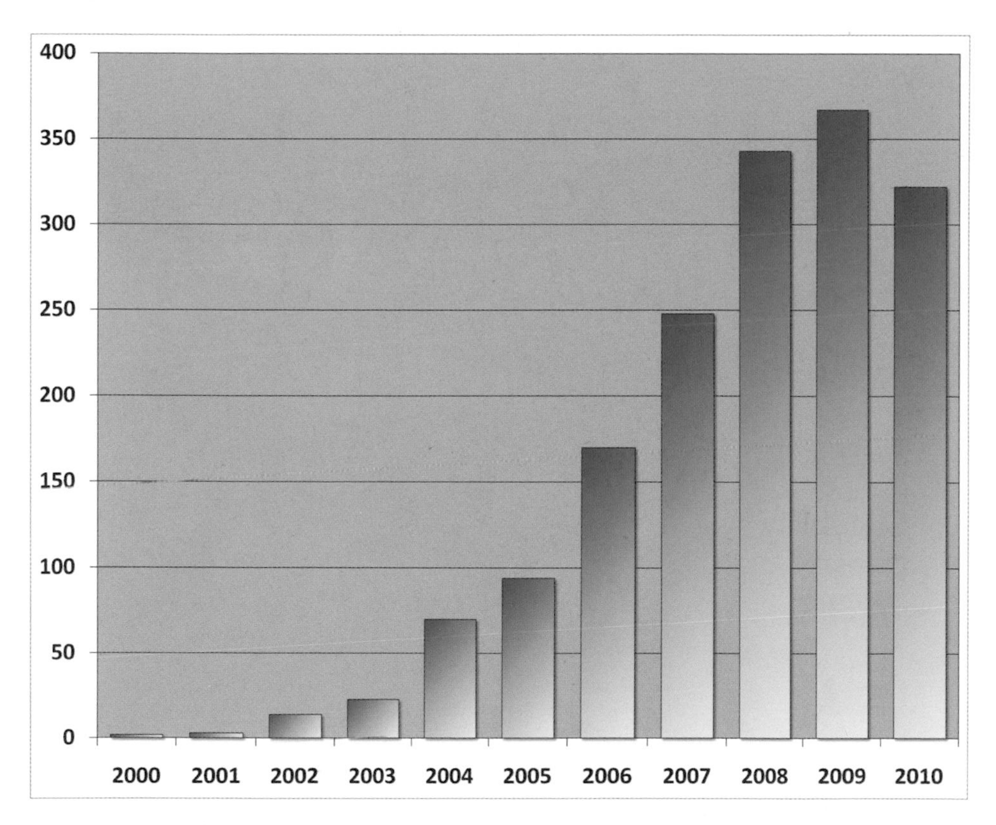

Figure 10. Graph showing the total number of Burmese pythons removed annually since 2000. Data courtesy of the National Park Service.

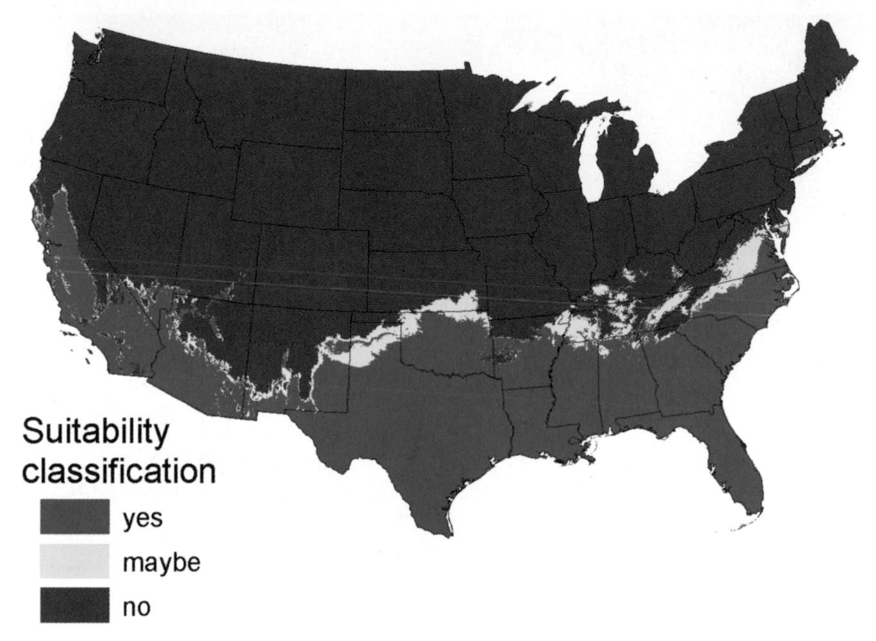

Figure 11. A 2008 analysis using average monthly temperature and rainfall resulted in this climate suitability map for Burmese pythons in the United States. Image courtesy of the United States Geological Survey.

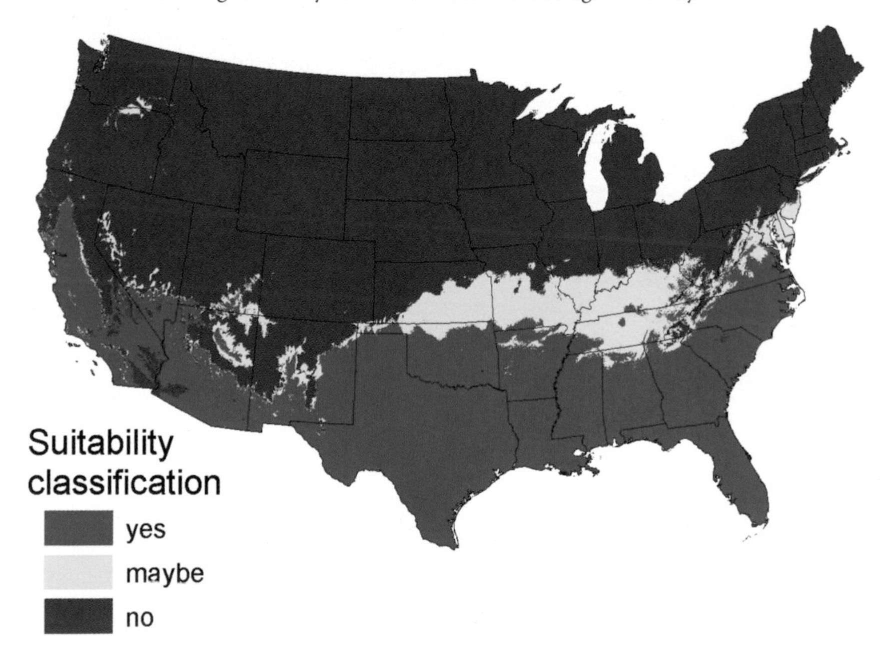

Figure 12. The 2008 study also postulated that, using IPCC projections for climate change in 2100, suitable climate for pythons may shift considerably northward over the coming century. Image courtesy of the United States Geological Survey.

Figure 13. In 2009 researchers used an ecological niche model to generate this map showing a very different projection for the potential expansion of pythons in the United States. Image courtesy of Pyron, Burbrink and Guiher.

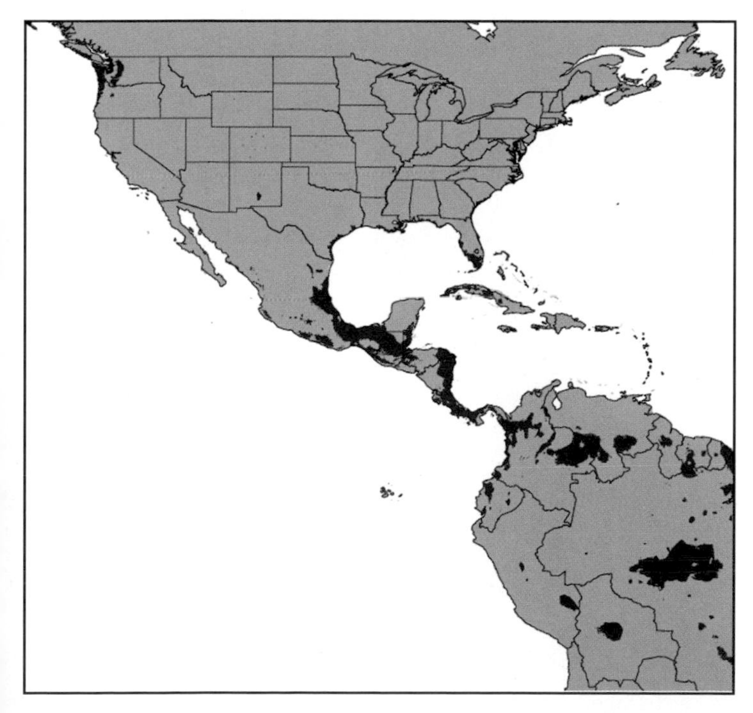

Figure 14. Applying IPCC projections for climate change in 2100 revealed little change in the potential expansion of pythons in the United States using an ecological niche model. Image courtesy of Pyron, Burbrink and Guiher.

Figure 15. During a severe cold snap in January 2010, rapid response teams conducting an assessment on the outskirts of western Miami located several large African rock pythons—including one measuring over 13 feet. Image courtesy of the National Park Service.

Figure 16. *Liberty and the Eagle,* by Italian sculptor Enrico Causici, presides to this day over Statuary Hall.

requirements prior to the enactment of the law, those who were in possession of a now-restricted snake or lizard were granted an exemption for the life of their pet. Those who had failed to file the proper paperwork were free to surrender the animals at any time. And commercial enterprises were still permitted to breed and export large constrictors elsewhere. Through the miracle of the Internet, Florida-bred snakes could still be sold as far away as Texas, California, or Seattle, or make a short journey across state lines into Georgia.

Still, what would prevent a return trip? Though state regulations were in place governing the transportation of wildlife into Florida, they were rarely enforced. As easily and efficiently as one might export predatory reptiles elsewhere, others might find their way back. After all, there were still plenty of large snakes available for purchase further north.

———

Much has been written about the relative impotence of current federal law against the introduction and establishment of nonnative species. A considerable amount of difficulty arises, for example, from the fractured manner in which nonnative species are addressed across multiple pieces of legislation and how that legislation is then implemented across various governmental agencies. For example, the National Invasive Species Act of 1996—despite its sweeping title—only concerns itself with the spread of aquatic nuisance species through the release of ballast water, and its provisions are administered jointly between the Secretary of Defense and the Secretary of Transportation. And while both the Animal Health Protection Act and the Public Health Service Act provide the Departments of Agriculture and Health and Human Services (respectively) with the authority to prohibit the importation of some foreign organisms into the United States, these powers are granted solely as protections against hazards to livestock, agricultural industry, and human health. Regulations on imports of foreign taxa, and the protection of terrestrial and aquatic ecosystems from their establishment and spread, fall largely under the purview of the Lacey Act as administered generally by the U.S. Fish and Wildlife Service.

The Lacey Act shares a history with the Everglades that has left an indelible impression to this day. The coastal reaches of south Florida are well known for the profusion of wading birds that once congregated there, in numbers so thick that naturalist John James Audubon once recounted how, in flight, they "blocked the light from the sun for some time." The birds would congregate in great communal rookeries as rains subsided and nearby marshes began to dry, building nests in which to rear their young for the season. Historically, hunters hearty enough to brave the biting swarms and tortuous waterways of the Everglades might view the cacophonous, stinking, guano-filled branches of a large nesting colony as the avian equivalent of El Dorado. At the turn of the twentieth century, the trade in feathers to adorn the fashionable hats worn by ladies was at a fever pitch, and a supply of delicate plumes could command a handsome price. A few hardscrabble hunters had been known to get lucky and amass personal fortunes virtually overnight.

To maximize profit, most would forgo picking off lone individuals in favor of finding one large rookery. The reasons were threefold. For one, dealers would pay top dollar for aigrettes—the light, wispy feathers exhibited by wading birds only during the breeding season. And rookeries, of course, would often contain hundreds—if not thousands—of birds to reap at one time. But most importantly, hunters targeted rookeries to capitalize on the birds' most vulnerable weakness: the indefatigable desire to protect their young. With every successive rifle shot, the birds would flap instinctively into the air, only to return a few seconds later to tend to their squawking progeny. Tethered to the trees by parental forces, an entire colony would be picked clean, leaving the future generation to die a slow death from starvation and exposure.

Not surprisingly, populations of wading birds plummeted precipitously, and depictions of the carnage left behind stoked the wrath of many. But the story unfolding deep in the Everglades backcountry was similar to stories unfolding elsewhere. Though once abundant, widespread hunting led to the disappearance of the carrier pigeon, and overexploitation was leading grouse and prairie chickens down a similar path. And while some states had begun taking measures to protect wildlife from wholesale slaughter, limited authority was available to prevent hunters from transporting animals killed in violation of state laws to other states where regulations were more favorable.

In addition to being an eight-term Republican Congressman, John F. Lacey was a staunch defender of wildlife. In particular, he lamented the damage being wrought to wild bird populations through market hunting. In 1900, he introduced a bill to Congress intended to empower the Secretary of Agriculture to better protect native wildlife through the regulation of trade across state lines. Passed later that same year and named eventually on his behalf, the law criminalized the interstate transport of any wildlife killed in violation of state law, assigned labeling requirements to shipments, and empowered states to regulate the sale of all wildlife entering their borders according to their respective laws.

Recognizing how the introduction of some foreign species resulted in harm to both wild game and agricultural interests, Lacey also included provisions to regulate the importation of such species under his law. Since its earliest iteration, several species were listed as "injurious" and explicitly prohibited from importation or passage across state lines, including mongoose, flying foxes, and starlings. This provision established a "black list" system of regulation, wherein all imported species are generally considered innocent of any harm and are permissible for trade until demonstrated otherwise, at which point they may be barred from further importation.

Though originally focused primarily on the protection of wild mammals and birds, the Lacey Act has been amended over time to extend protection to mollusks, crustaceans, fish, amphibians, and reptiles. The list of injurious species has also gradually expanded to prohibit trade in additional problematic animals. It has been noted by some that the Lacey Act has slowly evolved from a legislative tool against domestic poaching to an increasingly important control on the international trade of wildlife and the introduction of nonnative species.

But despite the noble intentions behind its origin, the Lacey Act is today a dated tool whose effectiveness against the onslaught of introduced species has been eroded slowly by the passage of time. After carrying the force of law for over 110 years, and despite a litany of notable impacts, a total of only

20 taxa—and their closest relatives—have since been listed as injurious to native wildlife and ecosystems, and thereby prohibited from import. To even casual observers—many of whom could likely rattle off a much longer list of problematic nonnative species—this number seems incomprehensibly low. This is particularly evident considering invasion hotspots like Hawaii, southern California, and south Florida, where literally hundreds of foreign species have already proven themselves capable of spiraling out of control.

Published assessments have helped illuminate the overall failures of the Lacey Act in preventing the introduction of nonnative species. Some have observed that the standard of guilt for listing species as injurious is far too high—the burden of proof is often met only *after* a given organism has successfully escaped, become established, and caused some measure of harm. Additionally, the lack of a specified time limitation on the process of listing an organism is a major shortcoming. Currently, the mean time between the petitioned listing of a species and the time in which a successful prohibition takes effect hovers just over *four years*, with some cases lasting as long as seven. During this delay, potentially dangerous organisms can continue to be imported legally and traded across state lines, and establish or expand populations freely.

The relative inertia that plagues the process of listing species as injurious is most troubling in our current age of global trade. Since 1991, the annual volume of live animal imports has roughly doubled. A 2007 study by Defenders of Wildlife used U.S. Fish and Wildlife Service records to quantify the number of live animal imports to the United States between 2000 and 2004. During this five-year period, well over *one billion* inventoried animals were legally brought to our shores—the vast majority of which were never identified to the species level. Of the roughly 6% that were, over 2,240 foreign species were recorded, though the actual number of nonnative species imported is likely much higher. The cumbersome, multi-year process currently necessary to prohibit importation of a single harmful species under the Lacey Act seems woefully inadequate for a country that averages imports of nearly 600,000 live animals *every day*.

It has been recognized that biological invasions typically involve four steps: transport of an organism from its point of origin, introduction to a foreign locale, establishment of a population, and a gradual spread beyond the initial point of introduction. It has further been noted that the Lacey Act, though arguably successful at impeding the transport of a select few organisms, has failed to provide the necessary authority to intervene in subsequent stages of invasion. The law provides no penalties for the intentional or accidental release of listed species, does not prohibit the ownership of such organisms, and does not appropriate funds towards the containment or eradication of established populations.

Yet for all its limitations and failures, the Lacey Act remains a key legislative tool in the protection of native ecosystems. As the full reach and impact of Burmese pythons became clearer, some began to vigorously lobby for the listing of pythons as an injurious species to help stem the tide of imports and possible new introductions. And by far, the loudest voice among them was that of Florida Senator Bill Nelson.

On February 3, 2009, Senator Nelson introduced S. 373, "a bill to amend title 18, United States Code, to include constrictor snakes of the species Python genera as an injurious animal." The original intent of the proposed bill was to close the spigot on international and interstate trade in all seven species of python, a list that included several commonly sold in the pet trade but not known to have established populations in the United States. Many prominent environmental and animal rights organizations cited "sufficient scientific evidence" as ample justification to sidestep the painfully long listing process in favor of proactive legislation. "An ounce of prevention is worth a pound of cure," read a joint statement issued by twelve organizations, including The Humane Society of the United States, Defenders of Wildlife, Sierra Club, and The Nature Conservancy. Not surprisingly, the move drew sharp criticism from owners and interests in the pet industry.

The introduction of S. 373 garnered a great deal of news coverage from media outlets. Within days, the Internet seemed ablaze with personal commentary from reptile owners, animal breeders, snake enthusiasts, and industry lobbyists. Leading the reactionary horde was the Pet Industry Joint

Advisory Council (PIJAC), an advocacy group that represents the interests of retailers, distributors, and suppliers for the live animal industry. Based in Washington, D.C., PIJAC bills itself as a service-oriented organization that "promotes responsible pet ownership and animal welfare, fosters environmental stewardship, and ensures the availability of pets"—though, some might argue, not necessarily in that order. When, as was certainly the case in south Florida, these three objectives found themselves in conflict, it seemed evident that keeping pets available was PIJAC's top priority. Neither the uncontrolled proliferation of thousands of constrictors in the Everglades, nor their necessary extermination by the hundreds, was sufficient to earn PIJAC's support for Senator Nelson's broad-reaching bill.

PIJAC and other critics blasted the law as an attempt to circumvent the role of scientific investigation, risk analysis, and public comment used procedurally by the U.S. Fish and Wildlife Service in listing injurious species. Shortly following introduction of the legislation, the group issued a position statement providing talking points and encouraging grassroots activism against the listing. Despite agreement that the Burmese python population in south Florida "needs to be controlled, and if feasible, eradicated," and that the escape and "occasional release of pythons by their owners" contributed to the problem, PIJAC vehemently dismissed S. 373 as an unacceptable means of assisting in the management effort or preventing new invasions. Members were encouraged to voice their objection at every possible opportunity as the bill made its way between subcommittees.

In June of 2009, Representative Kendrick Meek, from Florida's 17th District, introduced a companion bill (H. R. 2811) in the House of Representatives. The following month, during a hearing before the House Judiciary Committee, representatives from both PIJAC and the United States Association of Reptile Keepers (USARK) successfully persuaded legislators to amend the bill to include only three of the original seven species—Burmese, northern African and southern African pythons. But what little sympathy the groups managed to muster was only to be enjoyed ever so briefly.

The asphyxiation of Shaianna Hare in Florida earlier in the month brought about a vociferous outcry of anti-snake chastisement, published widely as opinions and editorials in local papers. In Washington, Senator Nelson continued his relentless campaign to list the species. Lifting high the unfurled skin of a 17-foot Burmese python on Capitol Hill, he cautioned his fellow legislators, "It's just a matter of time before one of these snakes gets to a visitor in the Florida Everglades." And in Fort Collins, Colorado, a group of scientists from the U.S. Geological Survey made public a hefty risk assessment painting nine large constrictors as potentially serious threats to the environment.

By the end of 2009, S. 373 had been amended yet again in subcommittee. In its final iteration, it included all nine constrictors assessed by the USGS to be of "moderate or high risk" to the United States. If passed, the legislation would ban their importation and interstate transport as injurious species under the Lacey Act. And there was little evidence to indicate it wouldn't earn the necessary votes. Over the previous year, pythons had been the subject of intensive backcountry hunts, exhaustive scientific scrutiny, and serious criminal investigations—seemingly galvanizing public concern. Political will seemed strong to enact the measure but, ultimately, Congressional efforts fizzled following an unexpected announcement. On January 20, 2010, during a visit to wildlife inspectors at John F. Kennedy International Airport, Interior Secretary Ken Salazar declared that the U.S. Fish and Wildlife Service would pursue an expeditious listing of all nine large constrictors as injurious species. The proposed rule would be published in the Federal Register in early February, followed by a brief public comment period.

With large snakes almost certain to be deemed federal outlaws, demand for the hefty herps plummeted. "No one is willing to give me $10,000 for a snake when they think they may be added to an injurious species list," noted one Oklahoma breeder. While some decried the impending government intervention as an attack on their personal liberties, it seemed increasingly evident that objections were largely rooted in simple economics.

In the United States, reptiles have become a hot commodity. Trade in snakes, lizards, turtles, and crocodilians has experienced considerable growth in recent decades. Today, industry professionals estimate five million Americans keep, or have formerly kept, some type of reptile as a personal pet. Within US households it is believed over four million boas and pythons are currently kept in captivity and, of those, roughly 100,000 are believed to be Burmese or African pythons. Only a few years ago, it might have been unusual to know someone who claimed to have a snake for a pet. Today, it is unusual not to.

Americans clearly love to not only own reptiles, but also sell them. It has been estimated that 82% of the worldwide export and trade in captive-bred reptiles flows from our borders. Spurred by economic interest, the industry has made great strides in the successful husbandry of numerous species—many of which are today imperiled in their native habitats. This is particularly true of pythons, as all known species but one have been successfully bred by keepers in the United States. It is an achievement that has sometimes been called one of the most significant, privately funded contributions to long-term conservation. Thanks to this effort, most pythons presently sold or exported in the United States come from captive-bred stock, thereby reducing pressures on wild populations in their homeland.

At the forefront of this industry are thousands of enterprising individuals that have managed to turn what is often a lifelong interest and hobby into a profitable enterprise. The United States Association of Reptile Keepers, the industry's premier advocacy organization, counts 12,000 professional breeders and sellers in their ranks. Businesses can run the gamut from individuals working out of a spare bedroom, to large-scale warehouse operations that turn out tens of thousands of new animals every year. Many businesses choose to specialize in particular types of reptiles, and a good number of these focus squarely on snakes.

For many, breeding boas and pythons can be a rewarding and profitable endeavor. Buyers might expect to pay between $400 and $1,000 retail for the purchase of an average adult constrictor. But some snakes can fetch considerably more, especially if they have been bred to exhibit attractively aberrant mutations or patterns for which customers will gladly pay a premium. Hypomelanistic harlequin boas, piebald ball pythons, albino granite Burmese pythons—all are engineered artistic interpretations of the manner in which each species was originally tinted by Mother Nature. Producing a lucrative clutch of color morphs often requires a hefty up-front fee for the purchase of breeding stock. But breeders are often willing to take the gamble as the rarer the variant, the more buyers are willing to pony up for a purchase. Premium python breeds and color patterns can sell for anywhere between $40,000 and $75,000.

Ron Greenberg spent 30 years working for a fiberglass manufacturer in northern California. After retiring, he and his wife opened Ron's Reptiles—a specialty reptile breeding and sales firm—from their home in Chico, California. In the past they have paid as much as $10,000 to purchase individual snakes with rare color patterns and have invested considerable time and money maintaining their large-scale operation. As authorities continued to push forward with plans to list large constrictors as injurious species, people like Greenberg began to worry their business—and passion—might be in the crosshairs.

And Greenberg's story is scarcely unique. In Colorado Springs, Michelle Pearson operates Art in Scales, a small family-run business specializing in the production of exotic color variants of smaller pythons and boas. She, too, worries what tighter regulation might bring. Bill Parker is owner and operator of Feeder Mice Unlimited in Oroville, California. He estimates his company produces about 40,000 rodents a year, primarily for purchase by reptile keepers. In recent years, he has already seen his business decline by 30% and is concerned further regulation might put him out of work.

The same could be said for Chad Duggin, who breeds and sells colorful boa constrictors out of the garage of his home in Murfreesboro, Tennessee, or Jeremy Stone who does the same from his home in Lindon, Utah. And Florida businesses stand to lose as well. Chris McQuade, Zelph Ridgeway, and Jeff Hornsby all operate reptile breeding and sales outlets along the southwest Florida coast, and all are wary of governmental regulation. "If you give them seven species," cautioned Hornsby, "they are going to want ten more."

Adding insult to injury, many reptile breeders and enthusiasts perceive an unfair bias in attempts at regulation. After all, if the ultimate goal of regulation is stemming the impacts of invasive species, why was nothing being done to target the persistent occurrence of feral cats, the most widespread carnivore in the country and an insatiable predator of native birds, mammals, and reptiles? And if the issue was truly one about public safety, weren't there bigger fish to fry? Over the years, hadn't exponentially more people drowned in swimming pools, suffered death in automobile collisions, or been maimed and killed by aggressive dogs? Statistically, the risk posed to personal safety by keeping large constrictors was minimal by comparison. "Have we forgotten that anything can harm you?" lamented Norman Friedman of Fort Walton Beach, Florida. "If this ban is truly because these species can be harmful, then please supply every American with a plastic bubble to live in."

The push to designate large constrictors as injurious was significant for one overarching reason: the animals being proposed for listing were already commonly available as personal pets. Never before had such a widely owned animal been petitioned for prohibition. Regulating these species required treading upon new ground, and that could bring inadvertent and unforeseen ramifications. The sustainability of reptile-based businesses topped the list for many—especially in light of dynamics already in motion.

The financial crisis of 2008 negatively impacted a wide array of industries and professions—including reptile dealers. Amidst the crumbling of stock markets and financial institutions around the world, suppliers as far away as Indonesia saw a severe drop in demand for pythons. And in the United States, the domestic market suffered a second setback as states and federal officials pushed new legislation forward. In 2010, the *Wall Street Journal* declared a "Bear Market in Boas" as breeders lamented a precipitous drop in demand attributed to growing fears of impending regulation. Feeling the squeeze, some exporters began trying to develop alternative markets elsewhere, particularly in Asia.

And the passage of captive animal regulations elsewhere had already impacted the lives of some individuals and organizations. Mike Hopcraft, for example, had become quite famous throughout Abbotsford, British Columbia, as the Reptile Guy. For years, he operated an exotic animal rescue facility that specialized in placing unwanted reptiles in new homes. But following the tragic mauling by Gangus the tiger, the province passed stringent new prohibitions on the ownership, transportation, and public exhibition of exotic animals. Once in place, the regulations bankrupted the facility, forcing him to relocate the 100 snakes, lizards, and crocodilians then under his care and, ultimately, driving Hopcraft out of his own home.

An outright ban on trade in large constrictors would almost certainly have some measurable effect on the profits and sustainability of some individuals and small businesses. But whether or not industry could adapt to such change remains the bigger question. Are breeders—like Sisyphus—condemned to forever push the same heavy boulder up the same steep hill? Were these otherwise industrious and creative entrepreneurs predestined to continue rearing pythons and mice even as regulation pushed them over the ledge of financial ruin? To the contrary, the pet industry has always demonstrated a keen awareness of market trends and consumer preferences and in the past has adapted to meet changing conditions. Consumer demand is, after all, the force that compels many to begin selling animals for profit, and it is demand that presumably would lead such entrepreneurs to other, more lucrative ventures. The powerful promise of revenue has lured computer programmers, tugboat operators, fiberglass plant managers, and simple hobbyists to abandon steady occupations in favor of starting their own husbandry ventures. And as demand waxes and wanes one could reasonably well predict that those same individuals would again rise to the challenge and adapt as necessary to stay profitable.

And fortunately, the animal kingdom is rich with interesting species beyond just large constrictors. Any number of new and novel organisms could be promoted and marketed for sale if the proposed regulation of pythons, boas, and anacondas ultimately proved successful. And thanks to the relative ease with which those species could be brought past our borders, the possibilities are really endless for the introduction of trendy new pets. Of course, the quest to bring increasingly exotic options to consumers would, without question, perpetuate a vicious cycle that invites new instances of biopollution.

———

Of the taxa currently prohibited from import under the Lacey Act, more than half were already present in the United States at the time of listing. Furthermore, many of these had already escaped from captivity and become established. Not surprisingly, analysis indicates the act has been most effective in preventing biological invasions when used proactively to discourage import. But blacklisting a species generally failed when employed as a reactionary response to animals already problematic or present. Thus effective prevention seems to require a perspective contrary to that from which the Lacey Act has functioned historically. Rather than leaving the borders wide open to a daily influx of half a million organisms presumed to do no harm, it seems necessary to assume the worst and permit entry to only those that can be reasonably demonstrated to do otherwise. In other words, presume all organisms are guilty of possible invasion until a thorough risk assessment says otherwise.

At its most basic, risk assessment entails carefully reviewing available scientific literature regarding a particular plant or animal. This analysis endeavors to identify whether or not favorable conditions might exist for the establishment and spread of the species following importation. Furthermore, risk assessments typically endeavor to identify the nature and extent of damage that might be expected to agriculture, human health, the economy, or the environment should invasion occur. The analysis is usually framed by examining the natural constraints that limit the occurrence of plants and animals in their native range (e.g. food availability, climate envelopes, habitat affinities) and comparing them to the character of the intended area of import.

Though the use of risk assessments is now generally viewed as an imperative for preventing future invasions, these assessments will always entail some element of uncertainty. At the moment, there remains a need to agree upon a set of best practices for their execution. Contrasting opinions persist as to what fundamental questions should be analyzed, the scale at which assessment should be made, and the manner in which captive-bred organisms should be evaluated. But even if such standards are ultimately established, it is still likely that risk assessments will never be totally foolproof. Past invasions have sometimes revealed how starkly the known life history of an organism can contrast with the unexpected habits it assumes in its introduced range—the biological manifestation of Jekyll and Hide. "What scientists thought was one serpent," writes Alan Burdick about the brown treesnake on Guam, "in effect was two: the native and the colonist, the preinvasive and the postlapsarian." And yet, despite its limitations, the governments of Australia, New Zealand, and Israel—in transitioning to their own respective "white list" systems—have all embraced pre-import screening as the single most effective means of minimizing the risk of invasion. Recent investigations also suggest that, if executed properly, the use of risk assessments can generate substantial economic returns—saving governments the costly expense of long-term management of invasive species.

In June of 2008, Congresswoman Madeleine Bordallo—representing the territory of Guam—took a bold step toward establishing a similar system for the United States by introducing the Nonnative Wildlife Invasion Prevention Act (H.R. 669) into Congress. As stated, the bill intended to "establish a risk assessment process to prevent the introduction into, and establishment in, the United States of nonnative wildlife species that will cause, or are likely to cause, economic or environmental harm or harm to other animal species' health or human health." The bill directed the Secretary of Interior to compile preliminary lists for both approved and unapproved species within the first three years following passage. The Secretary would also be afforded authority to quickly enact temporary emergency prohibitions in the face of potential harm from nonnative species. Once preliminary lists had been published, any interested person would be entitled to submit a proposal for the approval of additional species for import—though such proposals would also require that applicants pay a fee intended to cover all costs incurred by the government in conducting the necessary risk assessment. The bill also included specific provisions made for those legally possessing foreign wildlife prior to the enactment of the bill, which provided that those animals would be grandfathered and exempt from future regulation.

The broad reach of the proposed regulation elicited strong responses from interests well beyond those who had previously been arguing merely about south Florida's snake problem. A full spectrum of supporters and opponents offered their written and spoken opinions to the Subcommittee on Insular Affairs, Oceans, and Wildlife when its members gathered to deliberate on the bill in April of 2009. A coalition of biologists supported the policy as one analogous to those currently in place to screen food, drugs, and chemicals prior to their availability in the marketplace. A letter from the World Parrot Trust warned that passage of the bill would lead to the mass "liberation of large numbers of exotic animals into the wild." A joint statement issued by various animal welfare and conservation organizations strongly supported the measure but, likely in a bid to placate angry opposition, recommended extending exemptions to several small domesticated mammals already common in the pet trade. A member of the National Aquaculture Association testified that the bill would hamper the growth of the country's domestic seafood supply. One senior manager from the U.S. Fish and Wildlife Service expressed the agency's support of a new "white list" system for regulation, but also shared reservations about their abilities to administer and enforce the bill as written. And Marshal Meyers, the Chief Executive Officer of PIJAC, argued risk analysis, rather than risk assessment, would be preferable for pre-import screening, as the former can also be used to analyze important socio-economic and cultural considerations. "Congress must also carefully consider," he suggested, "both the financial costs and benefits of imported species."

Thanks perhaps to the rancor that the bill elicited, H.R. 669 never moved beyond committee. And yet, for long-time supporters of similar efforts, the defeat was far from surprising. After all, the idea of moving towards a white list approach in the United States is not new. The difficulties and deficiencies of administering the Lacey Act have been evident to the U.S. Fish and Wildlife Service for decades. In 1973, the agency boldly proposed banning the importation of all foreign animals except those deemed to pose a low risk of invasion. Citing the introduction of English starlings, giant toads, and walking catfish, the agency intended to correct policy shortcomings in a bid to prevent additional harm from introduced biota in the future. Reaction from breeders, industry lobbyists, and pet owners was then, as now, less than favorable. Though the agency labored to defend its proposal amidst a tempest of rumors, propaganda, and outright falsehoods, the effort was ultimately abandoned. But I can't help wonder: had the effort to overhaul the system nearly four decades ago proven successful, would the attractive photo of a Burmese python in the Florida Everglades still adorn my bathroom wall today?

10
The Unanswerable Questions

On a recent trip to Washington, D.C., my family and I enjoyed a tour of the United States Capitol. As the storied domicile of our nation's legislative branch, such tours are immensely popular with the public and are heavily attended. On the day we arrived, the visitor center was overflowing with a congested stream of wide-eyed people. After first being ushered into a nearby theater to enjoy a brief, but stirring, IMAX presentation, my family was invited to join a group of approximately 25 people for a walking tour. The Capitol, as it was explained to us, was still a place of business and, given the number of tours that occur there simultaneously, our guide was required to conduct the tour in hushed tones. He would use a microphone, however, and each of us was handed a set of headphones and a receiver to wear around our necks on a bright orange lanyard.

For nearly two hours, a portly man in a bright red jacket escorted us through the many rooms and passages within the domicile of our nation's legislative branch. All the while, speaking into his lapel, he regaled us with an encyclopedic knowledge of both the structure's history and the artwork that adorned it. There seemed no end to the narrative our guide could provide for every painting, sculpture, or architectural detail we encountered as we moved between the Rotunda and the Crypt below.

Our tour eventually led us to the Old Hall of the House, which served as the meeting space of U.S. Representatives for half a century before moving to the chamber in use today. This area is alternately known as Statuary Hall, as it now houses a collection of works commissioned by each of the individual states of the union to recognize the contributions of some of their most recognizable citizens. While standing amidst likenesses of Henry Clay, Jefferson Davis,

Daniel Webster, and Brigham Young, our guide endeavored to highlight the remarkable acoustics of the room. He asked us to remove our headphones and listen carefully; then he proceeded to distance himself from us some twenty paces. Our ears received his whispers with such clarity that I could not help but look upward to try and comprehend how sound waves played upon the ceiling above. It was then that my eyes caught sight of the giant snake hovering overhead.

At one time, *Liberty and the Eagle* presided over the desk of the Speaker of the House. Though the house chambers have moved, the ivory plaster sculpture remains in the same location today. Originally crafted by Italian sculptor Enrico Causici, the 13-foot-tall piece depicts three figures. A beautiful woman draped in robes holds a scroll in the center. A bald eagle is perched to her right with wings outstretched, and, to her left, a rough-scaled snake coils tightly around what appears to be a column. I found the snake to be surprisingly large—given the proportions of the piece it appeared to reflect a serpent of roughly eight feet or more.

It didn't take a scholar of American history to interpret the significance of either the eagle or the woman—they were clearly freedom and liberty personified. But, perhaps thanks to biases formed long ago during my Catholic upbringing, I found the presence of the snake somewhat odd, and its intended significance puzzling. And before I could ask for clarification, our guide ushered us hurriedly out of the chamber to make way for another group.

It was not until I arrived home several days later that I did a bit of research. It turns out that in sculpting the piece, Causici had a specific purpose for his imagery. What I had taken at first for a column was actually intended to be a fasces—a bundle of rods which, since Roman times, has become a traditional representation of power and jurisdiction. In positioning the snake coiled around the fasces, Causici intended to symbolize wisdom in governmental authority. But given the acrimony with which the regulation of snakes had been debated in recent years, I found utter irony in equating a large serpent with wise legislation. Still, it served to remind me just how differently every person regards large snakes, and how differently they might factor into one's world view. If Causici believed they were the harbingers of wisdom, who am I to argue?

No matter how carefully crafted, worded, executed, or interpreted, no single piece of environmental legislation will ever fully hit the mark to the complete satisfaction of any one person. Unlike the ink and paper with which bills are drafted, environmental issues are never black and white. They exist instead in a perennial state of gray—mired in personal beliefs about how we relate to the world around us and how we interpret our role in the larger biosphere. At their very core, environmental concerns are grounded in timeless questions of philosophy—questions which, though pondered and argued throughout history by great minds, are ultimately answered to the satisfaction of each only by his or her own proprietary reasoning. Put differently, every individual enjoys a deeply personal environmental ethic.

Imagine, for example, that you are gazing upon a mountain range in the distance. Half a dozen rocky crags loom high on the horizon before you, their snow-capped peaks obscured only slightly by a crown of wispy clouds. Their lower slopes are cloaked in a stand of greenery growing from the slow melting of waters from above. And in the valleys below, a handful of modest dwellings can be seen dotting the landscape. Consider the scene and ask yourself—what do you make of the sight before you?

Do you look upon the wooded tree line as a hopeful source of firewood for your family or as a potential wellspring of new medical discoveries? Of the many forms of life likely dwelling there, which are those worthy of scientific study, which are fair game to hunt, which can be cleared with impunity, and which are so sacred they should only be admired from afar? Would you share a prospector's inquisitive desire to learn what untold riches the mountains might hold, or dream instead of developing a vibrant community upon it? Would you look upon the valley dwellers with envy for enjoying daily life in such idyllic surroundings, or would you find their presence to be a blemish on an otherwise perfect representation of wilderness? Do you see the lofty peaks as a recreational challenge to be met with rope, knots, and axes, or are they worthy of untouched reverence as spiritual embodiments of the divine? Notions of how we should properly regard such spaces—and our sense of responsibility for them—are entirely subject to individual interpretation. Similarly, opinions about how we are entitled to enjoy, exploit, or interact with such resources are likely as numerous as the individuals with which we share space.

When viewed across populations and cultures, the multitude and diversity of opinions possible are not only staggering, but obviously pitted with seeds of strife. The deeply personal nature of environmental ethics—and the great dissonance in which they strive to coexist—manifests itself regularly wherever spaces are shared either physically or emotionally—from the densely populated regions of the California coast, to the largely uninhabited expanses of the Alaskan Arctic, to the humid lowlands of the Florida Everglades. The confluence of divergent personal philosophies fuels spirited debate on a wide array of environmental concerns—the reintroduction of gray wolves to Yellowstone, the modification or removal of dams in the arid American West, the development of offshore wind farms and solar arrays, the designation of large swaths of forest as protected wilderness, the most responsible manner in which to respond to a changing climate, and how best to respect the traditional knowledge and religious views of cultures that often hold elements of the natural world as sacred.

Not all environmental issues are equally subjected to acrimonious debate. There are some personal values—corresponding with Maslow's most basic of human motivations—that are fairly universally held. Though personal interpretations may be differently nuanced, the right of individuals to be afforded access to pure drinking water, safe foods, and clean air is generally appreciated and regarded as paramount. Thus, in a bid to guarantee our physiological needs and satisfy our desire for safety and well-being, environmental reforms addressing these issues—like the Clean Water Act, Clean Air Act, etc.—have generally enjoyed broad public support.

But when our shared resources become intertwined in the satisfaction of higher-level needs—the satisfaction of which not all people aspire to or are at liberty to enjoy—the environmental debate becomes far more contentious. By virtue of democracy and industry, the United States has long provided its citizenry both the wealth and security with which to pursue higher aspirations. Having calmed both our hunger pangs and our nerves, we have increasingly ascribed new meanings to our landscapes and resources beyond the satisfaction of our base needs. Today, our forests provide more than mere timber; they can also provide inspiration for artists, be a popular venue for family camping, or fulfill the promise of solitude and adventure. In our cities, open spaces are valued for their potential to provide a place to exercise the human body, the chance to picnic with friends, or the perfect setting for a romantic first date. Like our ancestors, we continue to hunt and fish, though today our motivations stem less from subsistence than from the desire to forge strong relationships with others, cultivate new skills, or escape the drudgery of everyday life. The very animals we pursue have themselves become more than mere food or beasts of burden—they have become personal companions, a catalyst for worldwide travel, and the focus of countless recreational hobbies.

The proliferation of Burmese pythons is clearly a pressing environmental concern in south Florida but, like other such issues, how it is interpreted hinges upon the subjective ethics of the observer. Furthermore, we enjoy sufficient luxury to heap layers of viewpoints upon it—formed largely by the meanings we ascribe to the organism itself. How we react to the proliferation of pythons is a function of whether we see the organism as a frightful beast, a family pet, a commodity, a skilled predator, a survivor, a creation of God, a manifestation of wildness, a trophy, an invasive pest, or a sentient being. Similarly, the meanings we ascribe to the organism inform our ideas about whether its intended purpose is to be captive, hunted, used for study, serve as an educational prop, fulfill some conservation goal, or simply be left alone as an equal resident of the biosphere.

As we ascribe deeper meanings to the natural world around us, it becomes the stage upon which many of us unfurl our perceived purpose. "What a man can be," Abraham Maslow writes succinctly, "he must be." At its greatest significance, the natural world can serve to help satisfy the ultimate motivation to "self-actualize," wherein we become that for which we are best fitted and become who we believe we are meant to be. Burmese pythons and their kin have been the catalyst for many observers to forge their identities and entrench themselves in their sense of purpose as journalists, hunters, scientists, commentators, breeders, hobbyists, businessmen, lobbyists, environmentalists, and entertainers. In the midst of such discordant views, it is important to remember that, ultimately, everyone is right.

Without question, people are inextricably linked to the planetary biosphere. Without the slightest bit of conscious intent, we are implicated in a broad spectrum of ecological processes that not only sustain our species, but in turn, sustain others. As we respire, we exchange elements of life with our atmosphere. As we grow, we assimilate organic elements into tissue and bone. We are hosts to parasites, vectors of disease, subjected to genetic mutation, and culled through natural selection. Throughout our lives we pollinate, migrate, extirpate, and procreate. Within our very bodies we host a diversity of flora and fauna that rivals the most verdant springtime garden. Tethered to all of Earth's life by a common metabolic thread, we interact as both predator and eventual prey. We are decomposers who, eventually, will in turn be decomposed. That we are of the natural world is indisputable and has been evident since our earliest ancestral record.

And so it is perhaps fitting to question when, in our collective history, we suddenly began viewing ourselves as standing apart from the natural system from which we originate. Today, our understanding of many environmental issues frames the workings of man as a manifestation of the unnatural. This is particularly true in the realm of exotic species, wherein the difference between that which is native and that which is not has classically been delineated by the actions of people. But at what point does mankind make the sizeable leap from organic to inorganic? At what point do we move beyond acting as an agent of nature and instead act counter to it?

In the New World, the advent of exotic species is usually tied to the arrival of Europeans from the east. Spanish explorers and conquistadors crossed the oceans with a traveling circus of plants and animals, introducing them to what is typically imagined as an isolated world. DeSoto brought pigs, Cortez brought horses, and with them all came disease. The introduction of these new species proved catastrophic for the Indian populations that received them, changed the landscape of the Americas forever, and in some cases, still continues to plague modern-day communities. And though we have traditionally been quick to assign blame for invasive species upon sea-faring intruders from the Old World, modern research suggests that the cultures they encountered had been acting as agents of introduction for thousands of years before European contact.

Mesoamerican populations are believed to have distributed tomatoes throughout South and Central America long before the arrival of Spanish explorers. Squash and many species of beans were similarly introduced across North America from their points of origin in the south. Peach palms originating in the Amazon basin were exported to Central America and eventually to islands in the Caribbean. And maize, having been first cultivated in the highlands of southern Mexico more than 6,000 years ago, was intentionally modified throughout subsequent millennia via hybridization and genetic manipulation to suit growing conditions across the Americas. Even herds of bison, commonly aligned with the great plains of the Midwest, are believed to have been exported eastward almost to the Atlantic coast by Native Americans through the intentional manipulation and clearing of forest land through fire. And yet, despite these conscious efforts of engineering and dispersal, few would interpret ancient man as a vector for nonnative species.

In comparing and contrasting the actions of ancient man against the capabilities of our contemporaries, it is easy to see how we differ in our abilities to thrust foreign organisms into new environments. Whereas our ancestors are believed to have taken millennia to colonize the Americas, we are now able to move swiftly around entire continents in a matter of hours. Modern populations dwarf in size and reach the known communities of our early ancestors. And perhaps most importantly, we have relatively recently learned to harness external sources of power, which greatly differentiates the scope of our impacts from those who were limited by the strength and stamina of bone and muscle.

Reviewing the march of human history presents some difficulties in clearly defining which species might be construed as foreign, nonnative, or exotic. Though our attempts to differentiate our own actions from that of the natural world are clearly influenced by social constructs, there is little argument that we are serving as a catalyst for an unprecedented pace of change in our biosphere. Fueled by explosive population growth, this change manifests itself as the unsustainable extraction of limited resources, the mass extinction of species in peril, the urbanization of planetary populations, and a rapidly warming climate. As a society, we struggle to forge from our disparate environmental views a vision for the scope and nature of change we are willing to accept. And in our quest to do the same for exotic species, some academics have fairly begun to question whether or not the unstoppable introduction of these new organisms is being unjustly maligned in the first place.

The basic tenets of invasive species biology find their origin in the work of Charles Elton, an English ecologist who published a work entitled *The Ecology of Invasions by Animals and Plants* in 1958. In the book, he provides the first comprehensive accounting of plant and animal invasions around the world. But more than just providing an inventory of problem pests, Elton's analysis introduces readers to foundational concepts that have dominated the field of invasive species management for well over 50 years. His work pioneered a niche-based theory of invasion—the idea that disturbed, simple ecosystems are more susceptible to introductions than stable environments of greater biological diversity. He successfully identifies the myriad pathways (mariculture, private collecting, hull fouling, deliberate introduction, the construction of canals, etc.) through which new species are introduced into terrestrial and marine environments, and outlined quarantine, eradication, and control as successive strategies for suppressing their occurrence. And

Elton passionately makes the case for preventing biological homogenization in the long term and preserving species richness whenever possible.

Whether wittingly or not, however, Elton used numerous metaphors in his text that would also go on to influence the tone and language with which biological invasions are described to this day. Within his first paragraph, he likens the rapid proliferation of invading species to a violent explosion, with the admonition that, "It is not just nuclear bombs and wars that threaten us." He goes on to describe the movement of exotic species as ". . . the quiet infiltration of commando forces, the surprise attacks, the successive waves of later reinforcements after the first spearhead fails to get a foothold, attack and counterattack, and the eventual expansion and occupation of territory from which they are unlikely to be ousted again." Elton even compares his writing to the work of a war correspondent in reporting on the outcomes of battle. Over subsequent decades, such militant and vitriolic rhetoric against all foreign species has become a hallmark of invasive species biology.

In his book *Out of Eden: An Odyssey of Ecological Invasion*, Alan Burdick has elaborated upon the odd dichotomy that exists in the minds of some invasion biologists. Although intellectually distraught by the appearance of a successful invader, scientists may also begin to find themselves simultaneously admiring its prowess. "An introduced species may be a nuisance, even a menace," he writes, "but it is also, in a sense, a winner, and even a biologist can't help but be impressed, in a scholarly sense, by a winner's ability to survive." In recent years, a number of scholars have taken quiet admiration a step beyond and begun to sincerely question the breadth of the brushstrokes with which all exotic species have been colored. Foremost among them is Mark Davis of Malacaster College in Saint Paul, Minnesota. Davis has written extensively about the inadequacies of some beliefs commonly held in the field—like the niche-based theory of invasion—which have not fully been supported by real-world observations. And Davis has also been critical of the freedom with which unsupported ideas have become canonized over time, as well as the parlance with which many choose to communicate the many facets of the issue. Invasive species management, for example, is often cited as an imperative in preserving the health or integrity of native ecosystems. But in recognizing ecosystems as simply the sum of their parts and processes, Davis sees little grounds for labeling them as either sick or healthy. As such, he cautions against the use of loaded terminology, acknowledging that, "When someone is referring to a healthy ecosystem, what they are referring to is an ecosystem the way they want it to be."

Davis and others have also challenged the oft-repeated mantra that exotic species threaten native species with extinction. Outside the realm of islands, lakes, and other insular systems, there is little evidence to support the notion that nonnative species alone are capable of wiping out competing species. On the ground, the opposite seems more often true. The introduction of foreign organisms commonly fails to replace native species, and instead, serves to increase an area's overall biological diversity. And in some cases, they may even be complicit in providing an unexpected service to the recipient community or restoring ecosystem function.

Hemigrapsus sanguineus is a Lilliputian crustacean that, save for growing no larger than a quarter, looks like any other crab one might find along an Atlantic beach. But the Asian shore crab, as it is known, was first recorded on the northeastern coasts of the United States in 1988, likely introduced through the release of ballast water. Today the miniscule invader thrives on beaches as far north as Maine and as far south as the Carolinas. In 2010, researchers at Brown University published a study on the dynamics of this particular invasion. For more than a year, the team conducted surveys of Asian shore crab populations and native species richness over nearly 20 miles of cobblestone beaches in Rhode Island. Researchers found that the combined effects of native cordgrass and ribbed mussels not only facilitated the invasion of the Asian shore crab but, where they grew densely together, the crabs occurred in densities more than 100 times greater than those observed in their native range. Though the omnivorous diet of the crabs has long been thought to present a threat to native species, the researchers also noted an increase in native species richness in those same areas. One could conclude, therefore, that the introduction of the Asian shore crab has not only failed to cause significant harm to the system, but rather, has contributed to the local abundance and diversity of species along the shore.

Other studies indicate that nonnative species might not only have a negligible effect on native diversity but, rather, might lend a hand in maintaining the overall function of native ecosystems. In 2007, nonnative giant Aldabra tortoises (*Aldabrachelys gigantea*) were purposely introduced to "rewild" Ile aux Aigrettes, a small island off the coast of Mauritius in the Indian Ocean. Large endemic tortoises of the genus *Cylindraspis* once lumbered across the island, grazing upon native vegetation and dispersing seeds of the slow-growing ebony tree. Due to intense logging, entire forests of ebony disappeared, as did the tortoises themselves, which went extinct by the mid-1800s. Though a limited number of trees survived the onslaught, the catalysts for their proliferation did not. Researchers released the Aldabra tortoises as an ecological proxy for their long-lost relatives in a bid to reforest the island. In a report published in 2011, scientists not only proclaimed success, but called for more "reversible rewilding experiments to investigate whether extinct species interactions can be restored."

Though Elton viewed the constant "chess play" of species between continents as unstoppable, others have seen the repositioning of organisms as an inevitable force in the messy process of evolution. Calls to become more tolerant towards the guaranteed arrival of new species are growing in number, and leading some to rethink the practicality of our long-standing goal of preserving landscapes to reflect historic, point-in-time conditions. Elton might have also agreed: in advocating for the preservation of variety, he writes, "Provided the native species have their place, I see no reason why the reconstruction of communities to make them rich and interesting and stable should not include a careful selection of exotic forms, especially as many of these are in any case going to arrive in due course and occupy some niche."

This discussion is becoming increasingly important as the specter of climate change looms large upon the planet. How are we supposed to interpret the manner in which ecosystems and organisms respond to global change at the hand of man? As human-caused warming increasingly challenges the survival of plants and animals in their natural places of origin, will we view their migration to new locales as invasion? And where organisms are unable to adapt on their own, do we permit their extinction forever rather than introduce them elsewhere? As our world changes, so might our outlook on the arrival of new species. In looking back at the cumulative influence of pre-Columbian Indian civilizations in the Americas, author Charles Mann writes, "If there is a lesson it is that to think like the original inhabitants of these lands we should not set our sights on rebuilding an environment from the past but concentrate on shaping a world to live in for the future."

The concept of biodiversity can be defined multiple ways. Usually, the idea refers to the total count of different species in an area—a park, a state, a country, etc. But biodiversity isn't always a game of numbers. Some areas are said to be more diverse if they host a greater spectrum of organism types—a greater cross-section of the kingdoms, orders, and families of life currently known to science. In this regard, the cornucopia of biological curiosities that inhabits our oceans contributes to an overall diversity of life that rivals that of even the most prolific terrestrial system. And yet, biodiversity can also be viewed on a much smaller scale—measured as the amount of possible variation that exists within the genetic code of individual species.

Just as diversity can be defined across broad conceptual scales, species richness can also be evaluated across differing scales of space. In evaluating biodiversity, we might consider an area as small as a neighborhood park, or as large as an entire continent. The vantage point from which we assess our relative gain or loss of diversity over time—particularly as a function of biological invasions—is important.

That introduced species are resulting in a loss of biodiversity is a commonly held maxim. "Remarkably," write Dov Sax and Steven Gaines, researchers at the University of California, "this view has formed in spite of the fact that most exotic species are not known to have major detrimental effects on native biota and that many studies show no reductions in biodiversity." In

2003, Sax and Gaines published a paper that examined diversity trends across global, regional, and local scales. Thanks largely to habitat loss, species diversity has decreased on the planet, prompting the view that we are in the midst of a modern mass extinction event. But when viewed from smaller regional and local scales, evidence indicates either no significant change or a net increase in the diversity of most areas. Plant richness in the United States alone has increased by approximately 20%. Thus, while the composition of populations may have changed dramatically, the total number of species present in smaller areas is on the rise.

Writer Alan Burdick draws a clear distinction between what some might see as an everyday view of diversity versus a more academic accounting:

> Most people live small, local lives and are grateful for whatever manages to thrive in their arena; they live in an alpha-diversity world. Whereas beta diversity is visible only on a grand scale, requiring some effort to take in, it speaks to the traveler and the reader of travel books. Its appreciation is a kind of luxury, although perhaps no less valuable for being one, the traveler would say. The paradox of biological invasions, and one reason scientists have such difficulty articulating its hazards, is that to the average backyard viewer, the gross result appears to be ecological addition rather than subtraction.

This leaves us with several questions to ponder. In our accounting of diversity, do we resign ourselves to only examine raw numbers, or should we weigh the value of the species we lose in favor of our net gains? Furthermore, are we to favor greater global diversity—which is likely to be experienced and appreciated by the few—over greater regional diversity that would be enjoyed by so many more? That far-flung regions of the Earth are now being seeded with an identical stock of hardy organisms has long been an argument for the prevention of introductions. Many have lamented that during their travels to remote regions of the planet, they encounter an increasingly common portfolio of life—some of which herald from the very place the traveler calls home. "It is the biological equivalent," wrote one scientist, "of flying from

Seattle to Paris and going to Starbucks for your coffee." Some have gone so far as to dub our current evolutionary era the Homogecene. In truth, however, recipient environments are not becoming truly homogenous—rather they are assimilating transcontinental invaders into unique, novel ecosystems the likes of which can only be appreciated by those with the rare fortune to notice it at all. And whether or not these newly reorganized assemblages of life are preferable to the systems of old is again, as Mark Davis might argue, only truly of subjective importance.

In 2007, billionaire business mogul Sir Richard Branson purchased the 120-acre Caribbean island of Moskito—part of the British Virgin Islands—with a stated purpose of creating the "most ecologically friendly island in the world." Only four years later, Branson announced plans to introduce both ring-tailed and red-ruffed lemurs to the island from their native Madagascar. Through his charitable foundation, Branson intended to establish wild populations of the endangered primates and help guarantee their survival against the deforestation that threatens their native homeland.

Almost immediately, scientists, environmentalists, and politicians initiated a chorus of concerns over the possibility of escape to nearby islands, the transmission of harmful diseases, and the potential impacts the lemurs' omnivorous appetites might have upon native wildlife—most notably the endemic dwarf gecko, one of the world's rarest lizards. Though Branson modified his plans slightly in the face of mounting criticisms, he still intends to establish a captive breeding colony on Moskito and continues to enjoy approval from the islands' minister of natural resources.

Branson's efforts have likely landed under the public microscope not only for the grandeur of his plans, but also thanks to his own personal star power. But his underlying intent to introduce foreign wildlife to areas beyond their natural range for the purposes of conservation is echoed around the world daily. Nonnative species are routinely moved across continents and oceans to establish ex-situ communities to thwart the specter of extinction. Exotic organisms are transported via rail, plane, pickup truck, and cargo container to satisfy the demands of research initiatives that strive to better understand the biology and ecology of threatened species. And in more informal settings, the traffic in exotic wildlife is often justified as a necessary means to engage the public in conservation issues. This classic view builds on Dioum's famous quote: "We conserve only what we love, we will love only what we understand, and we will understand only what we are taught." Many passionate, well-informed individuals view captive wildlife as the critical hook for teaching otherwise distracted audiences to pay attention to issues of conservation, putting people face-to-face with organisms that would otherwise be outside their daily scope of concern. As for the animals themselves, they are martyrs who involuntarily lose their freedom for the assumed purpose of securing the greater good of their wild brethren elsewhere. And some believe they also serve a secondary purpose. The close proximity in which they exist with their human admirers provides an increasingly nature-deprived populace to again marvel at the wonder and wildness of life. In speaking recently about the unfounded need for regulation on trade in large constrictors, Jamie Reaser, the vice president of the Pet Industry Joint Advisory Council, told an inquiring reporter, "A lot of children spend hours in front of the TV and computer and have very little contact with nature. So I think it's very important children have the opportunity to learn about animals and the environment."

Many people might agree, myself included. In my formative years, I captured and kept many snakes and other reptiles—experience that stoked in me an appreciation for those creatures that persists today. Over the years, I've known a great many students, mentors, environmentalists, and friends who have maintained a healthy interest in reptiles—and have indulged their fascination through ownership. Gordon Rodda and Bob Reed, in prefacing their controversial risk assessment of large constrictors, beautifully describe this dynamic:

> We can testify to these snakes' attraction personally, as we both have kept giant pet constrictors. We can attest to these snakes' beauty, companionability, and educational value. The love of nature is often

> originally fostered in one's own arms, where close contact with living things engenders a connection not otherwise possible. And size does impress. Thus the social value of protecting native ecosystems must be weighed against the social value of fostering positive attitudes about the protection of nature through giant constrictor ownership.

Many have written about the need to reacquaint people with the natural realm with which they were once so close. Aldo Leopold both lobbied for, and personally demonstrated, the importance of experiential encounters with nature in cultivating what he called a "land ethic." Rachael Carson, mother of the modern environmental movement, wrote eloquently about the merits of nurturing an appreciation for the natural world as a means of attaining a balanced life. Her book *Sense of Wonder* chronicles her attempts to impart that gift to her nephew Roger by facilitating exciting discoveries about the wild forces and beings around him. And most recently, Richard Louv has cautioned parents against the dangers of sequestering their children indoors and exposing them to the rising epidemic of "nature deficit disorder." Tactile, sensory free play outdoors—so common only a few generations ago as our grandparents scampered through family gardens or built shaky lean-tos in the woods—has all but disappeared from the experience of American youth. The harsh manner in which we have forcefully amputated ourselves from the natural world around us is increasingly being understood to have unfortunate implications for our mental, spiritual, and physical well-being.

A deep-seeded longing to connect to the natural world seems hard-wired in the human psyche, and the need to indulge that desire is becoming increasingly clear. At question, though, is whether or not captive wildlife—themselves severed from the natural world from which they hail—are an appropriate surrogate for nature-based experience. To many, captive animals no more demonstrate wildness than a photograph of the sun conveys its searing heat. So often, though, large constrictors are used as props for the entertainment and education of students and visitors across a wide array of programs and institutions. And so it must be asked: is a glancing encounter with a hefty snake in a sterile classroom environment necessarily the only—or best—means of engagement? And it is equally fair to question whether or not such proxies are truly a necessary evil at all, especially in areas that still boast considerable wilderness or rich collections of wild flora and fauna.

In south Florida, where the Everglades still run rampant with free-ranging bears and panthers, over 300 species of birds, copious aggregations of alligators and crocodiles, and innumerable winged insects, and boast waters teeming with fish and turtles, are there not sufficient opportunities available for experiential encounters that trump the need to import and imprison additional species? In the mosaic of habitats present in the Everglades, are there not already countless conservation issues to be highlighted, learned from, and from which to derive inspiration? And in a state already crawling with roughly 80 species, subspecies, and varieties of native serpents, are there not sufficient prospects with which to engage students without the need to employ giant boas, pythons, and anacondas? If not, perhaps the conservation battle is already lost.

And finally, if it is our intention to encourage conservation and species protection through the forcible capture, movement, incarceration, and use of exotic organisms, then we must constantly reevaluate our progress to determine whether or not to continue the practice. In the case of the Burmese python, unfortunately, it appears we have failed on both counts. Thus far, our vigorous trade in the species has done little to preserve the animals in their native range. And in the New World, the introduction of pythons now threatens to undermine longstanding conservation efforts in the River of Grass. Perhaps similar lessons remain yet to be discovered on the island of Moskito.

Gazing upon Causici's marble serpent, I begin to recast my interpretation of the wisdom it symbolizes. As Americans, we unfailingly protect the rights of all parties to enjoy and voice their myriad ideas and views. Citizens are permitted the liberty to practice their beliefs, insofar as they do not impinge upon the rights of others to do the same. But, of course, when interests too frequently find themselves in competition, and personal interpretations repeatedly find themselves at a stalemate, there exists a need to arbitrate between the two for the greater good of the community. In our democratic society, this is accomplished through the crafting and enactment of carefully worded legislation. This process—which includes not only the framing of law, but also its periodic evaluation through judicial review—does far more than merely identify prohibited behaviors. Rather, it is in the course of wordsmithing and reviewing draft statues and bills that the full gamut of opinions and ideas represented by the governed are laid bare and evaluated on their merits. Because it endeavors to be transparent and inclusive, this process is often lengthy, tumultuous, and painful. And though it is laborious and difficult, this effort is ultimately worthwhile, as it eventually gives birth to a national vision that, though never wholly satisfactory to any one individual, returns the nation upon a path to progress.

As a nation, our collective actions and attitudes towards foreign organisms currently manifest themselves in a schizophrenic manner. We marvel at powerful predators showcased in local zoos, but fear them when they suddenly appear in our neighborhood parks. We display outrage and dismay when exotic pets turn lethal, yet remain reluctant to infringe upon the personal liberties of owners. We exert great amounts of effort and resources in service to the preservation of native ecosystems, yet do very little to safeguard against their invasion by plant and animal pests. And though we sometimes recognize the damage some organisms can cause, we too often turn a blind eye to more problematic species of our liking. Wisdom, however, is realizing that none of these opinions is necessarily wrong. But it is also necessary to recognize the need for our community to surmount this dichotomy in the interest of progress. The painful process unfolding in the halls of the U.S. Capitol is not something to be avoided. Rather, true wisdom is embracing the process as a necessary measure in eventually forging a national vision and setting us on a course forward.

11
The Cycle Continues

The bright red shotgun shell I recovered from the edge of an Everglades levee in 2010 still sits upright on my desk. I pick it up often and toy with it between my fingers as I ponder the fate of pythons in the Everglades. I think about the more than 1,700 snakes that have so far been removed from the area. Though a lucky few temporarily become involuntary subjects in ongoing research studies, all are eventually euthanized for lack of a better option. Many more have likely died quietly and unceremoniously across the vast acreage of south Florida beneath the tires of passing motorists, within the jaws of hungry alligators, or amidst the sharp disks of heavy tractors. Pythons have somehow managed to survive the hazards of their new home, but in doing so, a great many have succumbed to heavy-handed justice dispensed by hunters, fearful homeowners, and biologists. And many no doubt have suffered fatal consequences wrought by prolonged exposure to bitter cold. The introduction of Burmese pythons into the Everglades has perhaps proven as disastrous to the snakes themselves as to those native species upon which they feed or might displace.

For me, the brass-and-plastic discard I maneuver between my fingers is a potent reminder of the high cost of our involvement with certain forms of life. In the case of pythons, we are most certainly justified in our fascination with such an arresting denizen of the Old World. In recent decades, we have indulged this admiration through close personal contact and attempted domestication. And yet this very act, though borne of the best intentions, today has led us down an unsavory path that mandates we choose between them and other species equally deserving of respect and protection.

For Burmese pythons in the Everglades, the writing is clear upon the palace wall. Observers have long since resigned themselves to the realization they will continue to infest south Florida into the foreseeable future, requiring costly amounts of time and money in a bid for limited management. Nonetheless, it has also been hoped that the high-profile disaster unfolding in the Everglades would provide sufficient cause to prevent similar introductions in the future. Over the past ten years, the python saga has prompted lively debate not only about how they are affecting their new range, but also how their introduction could have been prevented in the first place. Pythons pulled from the wilds of south Florida routinely pay the ultimate price for our transgressions—but if we fail to apply the knowledge we gain from the effort, these sacrifices are largely made in vain. And yet despite more than a decade toiling to control the impact of large constrictors in one of the nation's most treasured national parks, one might be surprised to find the situation mired in a repetitive series of familiar events.

Despite claims of mass mortality among Burmese pythons in the Everglades following the 2010 cold snap, a total of 322 individuals were captured by year's end—amounting to only a ten percent drop since the year before. Pythons radiating from the Everglades wilderness continued to be encountered in the rural outskirts of urban centers, where they pursued goats, chickens, and family pets. And in 2011, amidst a record-setting drought that left most wetlands of the Everglades parched—pythons were being found in areas farther north than they had been previously encountered. As one south Florida land manager lamented, "They appear to prosper regardless of extreme conditions, whether it's cold, wet, dry, or hot." With the benefit of 20/20 hindsight, it seemed increasingly evident that, despite less than favorable conditions, pythons were proving far more resilient than some had initially surmised.

Scientists remain actively engaged in debate about the finer details of south Florida's Burmese python invasion. Papers continue to be published arguing the merits of methodologies previously employed for the prediction of range expansion, and have called into question the validity of more conservative estimates of habitat suitability. Experts continue to offer evidence supporting controversial theories about how the present population was ultimately introduced. And based on more recent observations, population estimates for pythons in the Everglades continue to be revisited and revised.

The ongoing proliferation and dispersal of pythons has brought about new experimental attempts at control, which include the evaluation of passive python traps and the use of trained tracking dogs. Though each presents possible tools for suppressing populations, neither appears to hold promise for large-scale application and control. The proverbial silver bullet, though much sought after, remains as elusive as the pythons themselves.

In Florida, despite the enactment of state laws intended to better regulate and restrict the personal possession of large constrictors, transgressions continue to occur that suggest the potential for new disasters remains unabated. In May of 2010, wildlife officers in Miami charged Miguel Ruitort with three misdemeanors for illegally selling reticulated pythons to buyers without the proper licenses. Wildlife officers charged David Beckett for the improper caging and illegal possession of a Burmese python after his seven-foot-long serpent was found in the parking lot of an apartment complex near his West Palm Beach home. And rogue pythons continue to make surprise appearances, like the 14-foot, 200-pound northern African python encountered by Loren Mell as he walked his dogs outside his apartment in Tarpon Springs, Florida.

Given the difficulty in enacting similar provisions in the past, it is perhaps not surprising that federal regulation of large constrictors is still in flux. A full *six years* since the original petition was filed, the effort to list the Burmese python as an injurious species under the Lacey Act remains unresolved. Like others before him, Representative Tom Rooney of Florida has again introduced a bill in the House—H.R. 511—aimed at expediting the process. Simultaneously, the U.S. Fish and Wildlife Service continues to work toward an administrative listing of the same species under the Lacey Act. The future of any such proposal remains questionable, and debate persists. The United States Association of Reptile Keepers—and pet interests in general—are likely to continue fighting any proposed regulation of what they defend as a "nontraditional form of agriculture."

In the absence of comprehensive legislation, large constrictor species continue to be privately owned in other areas of the country, occasionally making unexpected appearances after they are released or make a break. In June of 2011, a six-foot Burmese python was found on top of a garbage truck in Cincinnati, Ohio, after the driver emptied the garbage dumpster at an area fast food restaurant. The following month, a ten-foot albino python was captured in a chicken coop in south Texas, after eating a goat, a dog, and a rooster. And in Longview, Washington, three large constrictors broke free from the garage of the private home where they were kept. Responding officers returned both a six-foot Burmese python and a ten-foot boa constrictor to the owner, but confiscated an eight-foot green anaconda. And later that same month, hunters captured a feral, nine-foot boa constrictor in Waiawa Gulch on the island of Oahu. Despite long-standing state prohibitions against the private ownership of snakes, the boa was the third live snake captured on Oahu that year.

In Omaha, Nebraska, Cory Byrne's nine-foot pet boa constrictor coiled unexpectedly around his neck while he was showing it off one evening to a female friend. Though only 25 pounds in weight, the snake proved strong enough to fatally constrict him even while a roommate struggled to pry it off. Arriving responders were eventually able to extract Byrne, but the 34-year-old victim would later succumb to his injuries at a local hospital. The tragic incident was notable on two counts—it was the first known lethal attack by a pet snake in the state of Nebraska, and it proved to be the first documented case of a fatal attack by a boa constrictor. But sadly, in the larger picture, Byrne's death only served to add yet another entry to an expansive list of incidents involving large, captive snakes.

In and around south Florida, reproducing populations of entirely new unwanted invaders continue to be documented in the Everglades ecosystem. In Port St. Lucie, researchers recently discovered a thriving population of African five-lined skinks (*Trachylepis quinquetaeniata*) reproducing wildly in the vicinity of a facility formerly used by a reptile dealer. Large, carnivorous Nile monitors (*Varanus niloticus*)—long since established in Cape Coral along the southwest Florida coast—now appear to have established a second population in ritzy Palm Beach County. Near Florida City, two-foot-long Oustalet's chameleons (*Furcifer oustaleti*) now freely move about the shrubbery and trees surrounding the site of a former animal importer, where numerous juveniles and adults—including egg-bearing females—have since been collected. And in Homestead, just outside Everglades National Park, a newly discovered population of omnivorous, four-foot-long black and white tegus (*Tupinambis merianae*) have been readily consuming small birds, rodents, frogs, snails, anoles, crayfish, fruits, spiders, eggs, and all manner of insects. The tegus are the suspected progeny of adults released by a former animal dealer, and now seem well poised to invade the Florida Keys, where coastal waters are simultaneously becoming increasingly infested with an explosion of venomous lionfish. Two new herpetological inventories penned in 2011 report a new spike in reptile and amphibian invasions—a total of 56 foreign species now occur throughout the state.

South Florida's odd menagerie of exotic creatures continues to stoke public fascination. *Swamp Wars*, a reality show that follows a team of Miami-Dade Fire Rescue personnel as they respond to panicked calls about wayward wildlife, recently debuted on Animal Planet. In early 2011, the SyFy network unveiled the latest campy offering in their line of original B-movies: *Mega Python vs. Gatoroid*, during which over two million viewers were treated to what could only be called a cinematic disaster featuring former teen queen rivals Tiffany and Debbie Gibson. And following in a long line of National Geographic specials on the topic, the producers of *Snake Underworld* inexplicably brought the likes of rocker Henry Rollins to explore the swamps of Florida in search of pythons and answers.

The chaos of the media circus loosely mirrors the turmoil that continues to persist in reckoning with an area awash in foreign organisms. Efforts to restore the Everglades ecosystem continue into the present while, at the same time, exotic wildlife remains free to move across the border in the name of commerce. Huge sums of time and money are invested in the planning, design, and construction of expensive water management projects to shield the remnant Everglades from the demands and pollution brought by the burgeoning south

Florida population. And as that same community grows larger and more prosperous, it presents ever-increasing opportunities to undermine those restoration efforts through the introduction of new forms of biopollution. And where the most harmful fouling inevitably occurs, a heavy burden is borne by all for the costly management efforts necessary in perpetuity.

In charting a course in south Florida, it must be asked: do we continue to struggle in service to the idealized, historical Everglades landscape known by our predecessors, or do we instead adopt a new vision that accounts for our future realities? Do we bravely accept an unfamiliar ecosystem that, though different, yields greater biodiversity and more complexity? Doing so, it should be noted, would not necessarily represent an abdication of time, effort and money. "These arguments are not intended to promote any idea of complete laissez faire in the management of the ecosystems of the world," wrote Charles Elton, ". . . the outburst of human populations and the advances of technical power have put an end to any idea like that. The world's future has to be managed, but this management would not be just like a game of chess—more like steering a boat." Indeed, with our hands on the tiller in the waters of the Florida Everglades, we must finally decide which way we want to go.

End Notes

Chapter 1: Snakes on a Sawgrass Plain

Page 5, intact for the benefit and enjoyment (16 U.S.C. § 1 1916)

Page 6, Roughly 6,000 years ago (Lodge 2010)

Page 8, be permanently reserved as a wilderness (16 U.S.C. § 410 1934)

Page 11, over twenty billion dollars (United States Government Accountability Office 2007)

Page 12, individual specimens . . . were captured (Meshaka, Jr., Loftus and Steiner 2000)

Page 12, At nearly twelve feet in length (Everglades National Park 1979)

Page 13, two Burmese pythons basking (Meshaka, Jr., Butterfield and Hauge 2004)

Page 13, eleven pythons were removed (Snow, Krysko, et al. 2007)

Page 13, Burmese pythons are no longer juveniles (Bhupathy and Vijayan 1989)

Page 13, to be established in the Everglades (Meshaka, Jr., Loftus and Steiner 2000)

Page 13, ultimately enjoy substantial vindication (Ray W. Snow, pers. comm.)

Page 14, held tightly in its jaws (Morgan 2003)

Chapter 2: Getting Acquainted

Page 17, I started getting a little nervous (Barron 2006)

Page 17, conducting a necropsy in the field (Ray W. Snow, pers. comm.)

Page 18, amazing alligator-eating python (Morgan, It's Alien Versus Predator in Glades Creature Clash 2005)

Page 19, some scientists continue to argue (Barker and Barker, The Distribution of the Burmese Python, *Python molurus bivittatus* 2008) (Jacobs, Auliya and Bohme 2009) (Reed and Rodda 2009)

Page 20, Burmese pythons are generally encountered (Barker and Barker, The Distribution of the Burmese Python, *Python molurus bivittatus* 2008)

Page 20, one of the fastest growth rates (Reed and Rodda 2009)

Page 21, A strong, prehensile tail (Wall 1921)

Page 22, conspicuous pelvic spurs (Gillingham 1982)

Page 22, several dozen eggs at a time (Reed and Rodda 2009)

Page 22, single-digit clutches and those in excess of 100 (Wall 1921)

Page 22, a clutch every two to three years (Robert N. Reed, pers. comm.)

Page 22, she will often forgo all food (Van Mierop and Barnard 1976)

Page 22, generate heat for her unborn young (Barker, Murphy and Smith 1979) (Van Mierop and Barnard 1978) (Wall 1921)

Page 22, in need of further study (Reed and Rodda 2009)

Page 22, particularly with regards to courtship (Barker, Murphy and Smith 1979)

Page 22, relative impact of disease and parasites (Reed and Rodda 2009)

Page 23, through a process known as parthenogenesis (Groot, Bruins and Breeuwer 2003)

Page 23, can reach ages in excess of thirty years (Reed and Rodda 2009)

Page 23, oldest known captive Burmese pythons (Arave 2009)

Page 23, she was special to me (Arave 2009)

Page 23, information on social behavior (Barker, Murphy and Smith 1979)

Page 23, feeding response (Secor and Diamond 1995)

Page 23, energy efficiency (Cox and Secor 2007)

Page 23, effects of visual deprivation (Grace, et al. 2001)

Page 25, Turkey Point Nuclear Power Plant (Grimes 1993)

Page 26, the storm's role in the appearance (Donnelly 1993)

Page 26, into the howling atmosphere (Bilger 2009)

Page 26, revealed a close kinship among the animals (Collins and Freeman 2008)

Page 26, in keeping with a single, large-scale release (Graziani, Heflick and Cole 2009)

Chapter 3: Breaking the Chain

Page 28, it's fun (Clayton DeGayner, pers. comm.)

Page 29, is found nowhere else on earth (United States Fish and Wildlife Service 1999)

Page 29, fewer than 200 individuals (Winchester 2007)

Page 30, the southernmost woodrat on record (Joanne Potts, pers. comm.)

Page 31, first wild Burmese python discovered (Wadlow 2007)

Page 33, $137 billion every year (Pimental, et al. 2000)

Page 33, facilitated the extinction of (Lowe, Boudjelas and De Poorter 2000)

Page 33, prey heavily on nesting seabirds (E. I. Fritts 2007)

Page 33, in southwestern Puerto Rico (Engeman, Laborde, et al. 2010)

Page 33, have spread throughout Caribbean waters (Schneider 2010)

Page 33, as serious agricultural pests (Lowe, Boudjelas and De Poorter 2000)

Page 34, during nearly two hundred years (Martinez-Morales and Cuaron 1999)

Page 35, by sorely disenchanted owners (Herald Tribune Staff 2009)

Page 35, to create a steady supply (Somma 2009)

Page 35, established in North Wales (Natural England 2009)

Page 35, now ranges across the entire island (Quick, et al. 2005)

Page 35, 700 feral California kingsnakes (Noticias EFE 2011)

Page 35, for the first time in 1979 (Wilson and Porras 1983)

Page 36, has been introduced worldwide (Tennant 1997)

Page 36, nearly one hundred of them (Snow, Krysko, et al. 2007)

Page 36, established population in the area in 1992 (Dalrymple 1994)

Page 37, introduced to the island of Guam (United States Geological Survey 2005)

Page 37, creep slowly through tree canopies (Rodda, Fritts, et al. 1999)

Page 37, densities that sometimes rivaled (United States Geological Survey 2005)

Page 38, innocence that betrays such island residents (Burdick 2005)

Page 38, and soiled feminine hygiene products (Rodda, Fritts, et al. 1999)

Page 38, silenced in Guam's inland forests (Fritts and Rodda 1998)

Page 38, a wider range of wildlife along the shore (Wiles, et al. 2003)

Page 38, a continued stream of prey (Fritts and Rodda 1998)

Page 38, an easy environment for foreign insect invasions (Fritts and Rodda 1998)

Page 39, young victims can require hospitalization (Fritts and McCloid 1999)

Page 39, to feed upon young domestic stock and fowl (Rodda, Fritts, et al. 1999)

Page 39, cause frequent power outages (Fritts and Chiszar 1999)

Page 39, a constant state of high alert (Fritts, McCoid and Gomez 1999)

Page 39, expensive barriers (Yoon 1992)

Page 39, Jack Russell terriers are employed (Engeman, et al. 1998)

Page 39, hovers around $5.6 million annually (Pimental, et al. 2000)

Page 39, it is difficult to identify (Rodda, Fritts, et al. 1999)

Page 41, just told people what was going on (Ray W. Snow, pers. comm.)

Page 44, clutches ranging from 21–85 eggs (Krysko, Nifong, et al. 2008)

Page 44, the stomach and intestinal contents (Snow, Brien, et al. 2007)

Page 44, the list of potential prey (Ray W. Snow, pers. comm.)

Page 44, 25 species in all (Dove, et al. 2011)

Page 45, by introduced predators on Guam (Dove, et al. 2011)

Page 45, roughly six feet in length (Ray W. Snow, pers. comm.)

Page 45, most warm-blooded terrestrial vertebrates (Reed and Rodda 2009)

Page 45, 30–40% of their prey (Cox and Secor 2007)

Page 45, utilize as little as 4% (Cox and Secor 2007)

Page 46, continue this practice in the New World (Ray W. Snow, pers. comm.)

Page 46, dozens of other vertebrates (Wilson, Mushinsky and Fischer 1997)

Page 47, entrance to a gopher tortoise burrow (Staats, Officials Hope Hunters, Public Help Them Root Out, Hunt Down Invading Pythons 2010)

Page 47, by abandoning their nests altogether (Ray W. Snow, pers. comm.)

Page 47, or perhaps as a deterrent (Shwiff, et al. 2010)

Page 48, a scary proposition (Morgan, Scientists Brace for Snake Invasion 2005)

Chapter 4: One of the Many

Page 49, the zoo was moved (Zoo Miami 2006)

Page 52, cryptogenic (Burdick 2005)

Page 54, to mainland Florida sometime in the 1940s (Lee 1985)

Page 54, effectively diluted the distinguishing characteristics (Bartlett and Bartlett 1999)

Page 55, the first nonnative reptiles in south Florida (Stejneger 1922)

Page 55, subsequent survey over forty years later (King and Krakauer 1966)

Page 55, conducted in 1983 (Wilson and Porras 1983)

Page 55, 40 introduced reptiles and amphibians (Meshaka, Jr., Butterfield and Hauge 2004)

Page 55, the Cuban treefrog (Meshaka, Jr. 2001)

Page 56, three possible avenues for introduction (King and Krakauer 1966)

Page 56, 56% of established species (Wilson and Porras 1983)

Page 57, The plague of toxic cane toads (King and Krakauer 1966)

Page 57, the University of Miami student (Wilson and Porras 1983)

Page 57, west coast community of Cape Coral (Enge, et al. 2004)

Page 57, could yield a variety of impacts (Wilson and Porras 1983)

Page 58, during the 80s and 90s (Acevedo 1999) (Associated Press 1989) (Miami Herald Staff 1982) (S. Hughes 1985) (Power 1990) (Miami Herald Staff 1992) (Medzerian 1988)

Page 58, soon Burmese pythons, reticulated pythons, etc. (Faiola 1991) (Spring 1983) (Hambleton and Donnelly 1991)

Page 58, one 100-pound python (Vernon 1993) (Plunkett 1993)

Page 58, they were being killed by hunters (Jeffrey 1989) (Arnold and Lyons 1983)

Page 58, on someone's back porch (Acevedo 1999) (S. Hughes 1985)

Page 58, frequency with which such encounters (Associated Press 1988)

Page 58, onto an active bulldozer (Hancock 1989)

Page 59, beneath a storage shed (Hecker 1991)

Page 59, at a popular Miami beach (Markowitz 1992)

Page 59, dozens of claims of ownership (Woodlee 1986)

Page 59, veteran wildlife trapper Todd Hardwick (R. Jones 1989)

Page 59, The most significant potential problem (Dalrymple 1994)

Page 60, he just picked the wrong species (Ray W. Snow, pers. comm.)

Chapter 5: Trial by Fire

Page 61, ten Burmese pythons made a lengthy trek (Dorcas, Willson and Gibbons 2010)

Page 61, in a mixed environment comprised of (Stevenson 2009)

Page 62, an invasive snake management and response workshop (Mazzotti, Brien, et al. 2007)

Page 63, as tiny and delicate as bats (Trousdale and Beckett 2005)

Page 63, beetles (Hedin and Ranius 2002)

Page 63, dragonflies (Wikelski, et al. 2006)

Page 63, locations of Florida panthers (Maehr, et al. 2002)

Page 63, American alligators (Goodwin and Marion 1979)

Page 63, and white-tailed deer (Labisky, Miller and Hartless 1999)

Page 63, emitted by tagged Florida manatees (Stith, Slone and Reid 2006)

Page 63, Everglades Snail Kites (Bennetts and Kitchens 1996)

Page 63, and White-crowned Pigeons (Strong and Bancroft 1994)

Page 64, instructions on the appropriate dispensation of anesthesia (Reinert and Cundall 1982)

Page 64, four pythons were released into (Mazzotti, Brien, et al. 2007)

Page 66, might also provide new avenues (Mazzotti, Brien, et al. 2007)

Page 66, in the midst of the brackish-water mangrove forests (Staats, 11-foot Python Captured at Rookery Bay 2010)

Page 66, in the ritzy enclave of Marco Island (J. Taylor 2010)

Page 66, half a dozen additional nests were also found (Ray W. Snow, pers. comm.)

Page 66, the waters of the Indian River Lagoon (UPI 2005)

Page 66, an 11-year-old boy found it slithering along (Morelli, 200-pound Snake Killed in Okeechobee Adds to Mounting Python Woes 2009)

Page 67, and even 200,000 (Morelli, Wildlife Experts Question Python Numbers in Everglades 2009)

Page 67, I believe it's probably around 1,000 (Morelli, Wildlife Experts Question Python Numbers in Everglades 2009)

Page 68, practicality of using unmanned aerial vehicles (G. P. Jones 2003)

Page 69, depending on landscape variables and animal movement (Reed and Rodda 2009)

Page 69, a telescoping boom and bucket truck (O'Conner 2009)

Page 70, a paper that seemed to shed some light (Rodda, Jarnevich and Reed 2008)

Page 71, several degrees over the next hundred years (IPCC 2007)

Page 72, a greater immediate threat than sea level rise (Sword 2009)

Page 72, began pushing for permit systems (Parker 2010) (Tompkins 2008)

Page 72, one watch list in the United Kingdom (Natural England 2009)

Page 72, one of his own captive ball pythons (Stevenson 2009)

Page 72, as an instance of "ecoterrorism" (Barker and Barker, Comments on a Flawed Herpetological Paper and an Improper and Damaging News Release from a Government Agency 2008)

Page 73, one of the largest and most diverse collections (Barker and Barker 2006)

Page 73, thirty miles north of San Antonio, Texas (Barker and Barker 2006)

Page 73, equally plausible to blame biologists (Barker and Barker, An Open Letter to Dr. Ray W. Snow 2008)

Page 73, as a precautionary measure against future invasions (Barker and Barker, The Precautionary Principle and Pythons 2010)

Page 73, engaged in a prolonged back-and-forth (Barker and Barker, Comments on a Flawed Herpetological Paper and an Improper and Damaging News Release from a Government Agency 2008) (Barker and Barker 2009)

Page 73, that ecological niche modeling strongly contradicted (Pyron 2008)

Page 74, captured on the loose in both areas (Asbury 2010) (Associated Press 1984)

Page 75, the warmth of a South Carolina summer (Dorcas, Willson and Gibbons 2010)

Chapter 6: Cold Blooded

Page 77, arrived at the house of Shaianna Hare (America's Intelligence Wire 2009)

Page 78, with third degree murder, manslaughter, and child neglect (Colarossi 2009)

Page 78, Darnell and Hare guilty of all charges (Hudak, Jurors Find Couple Guilty of all Charges in Killer-python Case 2011)

Page 78, nothing but a single road-killed squirrel (Hudak, Pet Python Likely Starving When It Killed Sumter County Child, Documents Show 2011)

Page 79, Febby Katwa knows well (Inambao 2001)

Page 80, a 38-year-old woman was attacked and killed (America's Intelligence Wire 2003)

Page 80, incapacitated by the pain (Hiaasen 2007)

Page 81, Max was certainly the underdog (UPI 2006)

Page 81, and his nine-month-old pit bull (Goode 2001)

Page 82, dotted his entire countenance (The Record 2001)

Page 82, used a Taser to ward off (The Pittsburg Tribune-Review 2006)

Page 82, at the Tarpon Springs Aquarium (Gray 2006)

Page 83, saved pet store owner Teresa Rossiter (The Record 2008)

Page 83, asphyxiation by neck compression (UPI 2008)

Page 83, twelve fatalities in the United States (The Humane Society of the United States 2008)

Page 84, Patrick Von Allmen was killed in Kentucky (UPI 2006)

Page 84, Ted Drees was killed by a 14-foot pet python (UPI 2006)

Page 84, Brothers Grant and Lamar Williams (Herszenhorn 1996)

Page 84, 16-foot Burmese python in Missouri (The Humane Society of the United States 2008)

Page 84, killed in Colorado by Monty (The Denver Post 2002)

Page 84, Monty was euthanized (The Denver Post 2002)

Page 85, for his less-than-friendly disposition (Gauen 2000)

Page 85, coiled beneath a nearby sofa (Goodrich, Parents Knew Snake That Killed Child Was Dangerous, Prosecutor Charges - Death Was an "Unforeseeable and Horrific Tragedy," Defense Says 2000)

Page 85, faced a maximum penalty of ten years in prison (Gauen 2000)

Page 85, knowingly placed the child in harm's way (McClellan 2000)

Page 85, you have the responsibility to find out (Goodrich, Couple Whose Son Was Killed by Pet Snake Believes a Jury Will Acquit Them - Centralia, Ill., Parents Have Received Some Hate Mail, Hostile Calls - Trial is Set For Nov. 15 1999)

Page 85, there could be set a bad legal precedent (Goodrich, Couple Had No Way To Know Snake Would Kill, Expert Says - Python Breeder Testifies in Trial Over Boy's Death 2000)

Page 86, Derek Romero was crushed to death (The New York Times 1993)

Page 86, 8-foot python escaped from his enclosure (The Humane Society of the United States 2008)

Page 86, by her father's 8-foot reticulated python (The Humane Society of the United States 2008)

Page 86, 8-year-old Amber Mountain (Associated Press 2001)

Page 87, accounted for 32 human fatalities (DogsBite.org 2010)

Page 87, between 11 and 33 deaths annually (Langley 2009)

Chapter 7: The Best Hope

Page 89, from an exotic pet breeding facility on Grassy Key (Engeman, Woolard, et al. 2006)

Page 90, 79 people would eventually be diagnosed (Fiala 2005)

Page 90, ban on the importation of (Wahlberg 2003)

Page 90, a dangerous threat to Florida's ecology (Engeman, Woolard, et al. 2006)

Page 90, confirmation of the population on Grassy Key (Perry, et al. 2006)

Page 91, had not yet spread beyond Grassy and Crawl Keys (Engeman, Witmer, et al. 2007)

Page 91, approximately 200 individuals had been trapped (Clark 2007)

Page 91, officially declared the noxious pest eradicated (South Florida Water Management District 2011)

Page 92, ranging between $100 and $137 billion (Pimental, et al. 2000) (Strickland 2009) (Morgan, Burmese python tops list of invasive killers 2009)

Page 93, giant African snails were brought to Miami (Smith, Whilby and Derksen 2010)

Page 93, the animal was declared eradicated (Poucher 1975)

Page 93, at the end of the effort (Engeman, Constantine and Bunting 2007)

Page 93, few wading birds that readily prey on eggs and nestlings (Johnson and McGarrity 2009)

Page 94, pulled an unusual catch from a four-acre retention pond (Haas 2009)

Page 94, a third piranha floating among the stinking mass (Ferraro and Moody 2009)

Page 97, some suggested circulating recipes (Grimm 2009)

Page 97, entire cookbooks had been authored (J. Gorman 2011)

Page 97, form a strong disincentive to total eradication (Burdick 2005)

Page 98, bore "extraordinarily high levels" of mercury (Pittman 2009)

Page 98, might be routinely making meals of large, top predators (Kessler 2010)

Page 98, an incentive to entice hunters to help (Morgan, Florida's Pythons Could Have a Bounty on Their Head—and Body 2009)

Page 98, that he himself had been throwing over the perimeter wall (Tomb 1995)

Page 99, errors in identification (Somaweera, Somaweera and Shine 2010)

Page 100, favor with the recipient culture over time (Nunez and Simberloff 2005)

Page 100, a sanctioned "python posse" (Kam 2009)

Page 100, Governor Crist wants to take action (Morgan, The Python Posse Is Coming to the Everglades 2009)

Page 100, a nine-foot python off a boardwalk (Fleshler 2009)

Page 101, not even seeing signs of anything (Bryen 2009)

Page 101, managed to capture only 39 Burmese pythons (Florida Fish and Wildlife Conservation Commission 2009)

Page 101, a well-attended course on python-wrangling (Cocking, Pythons 101: Learning to Hunt Everglades Critters 2010)

Page 101, reminiscent of the Keystone Kops (Schall 2010)

Page 101, the number of pythons captured was certain (Fleshler, Everglades Hunt for Burmese Pythons Fails to Catch Any Snakes 2010)

Page 102, inquiring about where to spot pythons (Cave 2010)

Page 102, existence of a second species of python (Spinner 2009)

Chapter 8: The Big Chill

Page 103, six northern African pythons (Ray W. Snow, pers. comm.)

Page 103, strong evidence of a reproductive population (Reed and Rodda 2009)

Page 105, and a sole box turtle (Mike Rochford, pers. comm.)

Page 105, a larger risk assessment (Reed and Rodda 2009)

Page 105, we feel it is a misrepresentation (Jacobson, et al. 2009)

Page 105, a second group of academics (Bartelt, et al. 2010)

Page 106, USGS Associate Director for Biology publicly defended (Haseltine 2010)

Page 106, getting funding for injurious snake research (Barker and Barker, Climate-matching Predictions for Spread of Giant Snakes in U.S. "Grossly Exaggerated" 2010)

Page 106, the coldest 12-day period on record (National Weather Service 2010)

Page 107, 90,000 deceased snook (Morgan 2011)

Page 107, over 240 manatees were discovered (Florida Fish and Wildlife Conservation Commission 2010)

Page 108, unusually cold weather for the region (Dorcas, Willson and Gibbons 2010)

Page 108, relocated to an outdoor laboratory in Gainesville (Avery, et al. 2010)

Page 108, ten snakes were being tracked (Mazzotti, Cherkiss, et al. 2010)

Page109, we might have lost maybe half of the pythons (Fleshler and Huriash 2010)

Page 109, pythons can't exist more than a short period of time (Fleshler and Huriash 2010)

Page 111, he has captured over 300 pythons (Morgan, Python Hunter: Miami Man Takes on Glades Swamp Serpents 2009)

Page 113, likely the largest male ever (Reed, Giardina, et al. 2011)

Page 115, a twelve-foot green anaconda near death (Tampa Tribune Staff 2010)

Chapter 9: Legislation-Come-Lately

Page 118, well over 1 billion new animals (United States Government Accountability Office 2010)

Page 119, thirty-two-year-old Tania Dumstry-Soos was mauled (The Calgary Herald 2007)

Page 119, a 200-pound chimp named Travis (O'Connor 2009)

Page 119, in an attempt to repair the damage (O. Johnson 2011)

Page 119, dragged to the bottom of a deepwater tank (Kluger 2010) (Colarossi 2010)

Page 120, freed from its enclosure for a feeding (Sheerha 2010)

Page 120, famously offering paying customers (Phillips 2010)

Page 120, patas monkeys that successfully absconded (French 2010)

Page 120, the wayward tiger rattlesnake (Boone, Grant Park Toddler First to Encounter Deadly Snake 2010)

Page 120, the Egyptian cobra that went missing for a week (Batchelor 2011)

Page 121, Florida leads the country (Born Free USA 2011)

Page 121, loitering near an apartment complex in Tarpon Springs (United Press International 2011)

Page 121, planted his 14-foot Burmese python Sweetie (Morelli, Officials: Trapper Faked Capture of 14-foot Python in Bradenton 2009) (Florida Fish and Wildlife Conservation Commission 2009)

Page 121, uncaged, 11-foot long Burmese python roaming (Behnke, Large Python Seized from Crestview Resident's Home 2009)

Page 121, foments an unfair portrayal of captive animal attacks (Barker and Barker, Herpetoculture in the 21st Century 2010)

Page 122, from the suburbs of Chicago (Sullivan 2010)

Page 122, the state of Rhode Island (Rodrigues 2010)

Page 122, like those in Ravenswood, West Virginia (Hunt, Town Will Require Owners of Exotic Pets to Get License 2010)

Page 122, measures passed by the legislature of North Carolina (Kiley 2009)

Page 122, the Texas legislature passed measures (Tompkins 2008)

Page 122, governor Ted Strickland issued an executive order (Steinmetz 2011)

Page 122, authorities in British Columbia enacted (Cooper 2010)

Page 122, violators now face stiff penalties (Pearson 2009)

Page 123, author Alan Green reminds us why (Green 1999)

Page 123, Florida is often credited with enacting (Green 1999) (Hardin 2010)

Page 123, officials initiated a series of captive wildlife regulations (Hardin 2010)

Page 123, a tiered permit system that is somewhat unique (Environmental Law Institute 2010)

Page 124, permit holders for *Class I* and *Class II* wildlife (Florida Fish and Wildlife Conservation Commission 2011)

Page 124, overworked wildlife agencies are often forced to prioritize (Green 1999)

Page 124, propagule pressure (Kolar and Lodge 2001)

Page 125, the frequency and persistence of introduction (Burdick 2005)

Page 125, to personally possess any of these species (Florida Fish and Wildlife Conservation Commission 2008)

Page 126, issued permits for fewer than 400 (Samuels, Critters Capture Capitol Culture 2010) (Flemming 2010)

Page 126, criticized as being too reactive in nature (United States Congress Office of Technology Assessment 1993) (Samuels, State Poised to Ban the Sale of Burmese Pythons 2010) (Dolinski 2009) (Herald Tribune Staff 2009)

Page 126, many welcomed the new rules (Morelli 2010)

Page 126, organize a year-round amnesty program (Behnke, FWC Orders Amnesty Program for Reptiles of Concern 2009)

Page 126, continuing to allow for trade between breeders (Samuels, State Poised to Ban the Sale of Burmese Pythons 2010)

Page 127, others might find their way back (Kaufman 2011)

Page127, the fractured manner in which nonnative species are addressed (Jenkins 2007)

Page 129, passed later that same year and named eventually on his behalf (Anderson 1995)

Page XX, the Lacey Act has been amended over time (Anderson 1995)

Page 130, with some cases lasting as long as seven (Fowler 2007)

Page 130, annual volume of live animal imports has roughly doubled (Jenkins 2007)

Page 131, biological invasions typically involve four steps (Kolar and Lodge 2001)

Page 131, failed to provide the necessary authority to intervene (Green 1999) (Fowler 2007)

Page 131, cited "sufficient scientific evidence" as ample justification (Animal Welfare Institute 2009)

Page 132, bills itself as a service-oriented organization (Pet Industry Joint Advisory Council 2010)

Page 132, vehemently dismissed S. 373 (Pet Industry Joint Advisory Council 2009)

Page 132, include only three of the original seven species (Koss 2009)

Page 133, just a matter of time before one of these snakes gets to a visitor (Daly 2009)

Page 133, made public a hefty risk assessment (Reed and Rodda 2009)

Page 133, noted one Oklahoma breeder (Scheck 2010)

Page 134, estimate five million Americans (Wyatt 2009)

Page 134, roughly 100,000 are believed to be (Wyatt 2009)

Page 134, 82% of the worldwide export and trade (Wyatt 2009)

Page 134, one of the most significant, privately-funded contributions (Barker and Barker, On Burmese Pythons in the Everglades 2009)

Page 134, thereby reducing pressures on wild populations (Cerabino 2009)

Page 134, 12,000 professional breeders and sellers in their ranks (Kaufman 2011)

Page 135, to pay between $400 and $1,000 retail (Dorell 2010)

Page 135, can sell for anywhere between $40,000 and $75,000 (Scheck 2010) (Kaufman 2011)

Page 135, they have paid as much as $10,000 to purchase (Daugherty 2010)

Page 135, Art in Scales, a small family-run business (Dorell 2010)

Page 135, already seen his business decline by 30% (Scheck 2010)

Page 135, who breeds and sells colorful boa constrictors (Carey 2010)

Page 135, from his home in Lindon, Utah (Kaufman 2011)

Page 136, breeding and sales outlets along the southwest Florida coast (Engstrom 2010)

Page 136, an insatiable predator of native birds, mammals, and reptiles (Kaufman 2011) (Dauphine and Cooper 2011)

Page 136, please supply every American with a plastic bubble (Gibson 2010)

Page 136, severe drop in demand for pythons (Agence France-Presse 2008)

Page 136, breeders lamented a precipitous drop in demand (Scheck 2010)

Page 136, develop alternative markets elsewhere (Agence France-Presse 2008)

Page 137, the province passed stringent new prohibitions (Baker 2011)

Page 137, abandon steady occupations in favor of (Scheck 2010)

Page 138, generally failed when employed as a reactionary response (Fowler 2007)

Page 138, generally viewed as an imperative for preventing (Secretariat of the Convention on Biological Diversity 2010)

Page 139, the native and the colonist, the preinvasive and the postlapsarian (Burdick 2005)

Page 139, governments of Australia, New Zealand, and Israel (Global Invasive Species Programme 2008)

Page 139, can generate substantial economic returns (Springborn, Romagosa and Keller 2011)

Page 139, the bill intended to (Bordallo 2009)

Page 140, as one analogous to those currently in place (D. M. Lodge 2009)

Page 140, liberation of large numbers of exotic animals into the wild (Gilardi 2009)

Page 140, various animal welfare and conservation organizations (Defenders of Wildlife, et al. 2009)

Page 140, hamper the growth of the country's domestic seafood supply (Martin Jr. 2009)

Page 140, reservations about their abilities to administer (Frazer 2009)

Page 140, argued risk analysis, rather than risk assessment (Meyers 2009)

Page 141, in 1973, the agency boldly proposed (Fowler 2007)

Page 141, the agency intended to correct deficiencies (United States Fish and Wildlife Service 1974)

Page 141, was then, as now, less than favorable (Meyers 2009)

Chapter 10: The Unanswerable Questions

Page 143, wisdom in governmental authority (Architect of the Capitol 2011)

Page 145, Maslow's most basic of human motivations (Maslow 1943)

Page 148, DeSoto brought pigs (Mann 2006)

Page 148, the intentional manipulation and clearing of forest land (Mann 2006)

Page 150, by a winner's ability to survive (Burdick 2005)

Page 150, not fully been supported by real-world observations (Davis 2011)

Page 151, an ecosystem the way they want it to be (Breining 2010)

Page 151, little evidence to support the notion (Breining 2009)

Page 151, through the release of ballast water (Altieri, et al. 2010)

Page 152, in a bid to reforest the island (Hansen, et al. 2010)

Page 152, to investigate whether extinct species interactions can be restored (Griffiths, et al. 2011)

Page 152, the constant "chess play" of species (Elton 1958)

Page 152, in the messy process of evolution (Breining 2009)

Page 152, are in any case going to arrive (Elton 1958)

Page 153, concentrate on shaping a world to live in (Mann 2006)

Page 154, show no reductions in biodiversity (Sax and Gaines 2003)

Page 154, appears to be ecological addition rather than subtraction (Burdick 2005)

Page 155, going to Starbucks for your coffee (Bailey 2010)

Page 155, dub our current evolutionary era the Homogecene (Bilger 2009)

Page 155, most ecologically friendly island in the world (Harrison 2011)

Page 155, from the islands' minister of natural resources (Black 2011)

Page 156, A lot of children spend hours in front of the TV (Fleshler and Williams 2010)

Page 157, in cultivating what he called a "land ethic" (Leopold 1949)

Page 157, chronicles her attempts to impart that gift (Carson 1956)

Page 157, for our mental, spiritual, and physical well-being (Louv 2008)

Chapter 11: The Cycle Continues

Page 161, only a ten percent drop (Morgan 2011)

Page 161, goats, chickens, and family pets (Trischitta 2011)

Page 161, appear to prosper regardless of extreme conditions (South Florida Water Management District 2011)

Page 161, merits of previously published methodologies (Rodda, Jarnevich and Reed 2011)

Page 162, present population was ultimately introduced (Willson, Dorcas and Snow 2010)

Page 162, continue to be revisited and revised (Reed, Hart, et al. 2011)

Page 162, evaluation of passive python traps (Reed, Hart, et al. 2011)

Page 162, illegally selling reticulated pythons (Ferraro 2010)

Page 162, wildlife officers charged David Beckett (Fleshler 2011)

Page 162, northern African python encountered by Loren Mell (Martinez 2011)

Page 163, nontraditional form of agriculture (Snyder 2011)

Page 163, on top of a garbage truck (Associated Press 2011)

Page 163, captured in a chicken coop in south Texas (KHOU 2011)

Page 163, three large constrictors broke free (Lystra 2011)

Page 163, in Waiawa Gulch on the island of Oahu (Jinbo 2011)

Page 163, would later succumb to his injuries (M. A. Beck 2010)

Page 163, first documented case of a fatal attack (Cole and Nelson 2010)

Page 163, thriving population of African five-lined skinks (Krysko, Johnson, et al. 2010)

Page 164, established a second population in ritzy Palm Beach County (Nolin 2011)

Page 164, two-foot-long Oustalet's chameleons (Gillette, et al. 2010)

Page 164, omnivorous, four-foot-long black and white tegus (C. Hughes 2010)

Page 164, the suspected progeny of adults released (Steele 2010)

Page 164, a new spike in reptile and amphibian invasions (Meshaka, Jr. 2011) (Krysko, et al. 2011)

Page 164, SyFy network unveiled the latest campy offering (B. Gorman 2011)

Page 165, more like steering a boat (Elton 1958)

Works Cited

Acevedo, Pedro. "Jaunty Python 9-foot Snake Sneaks Out of Cage," *The Miami Herald*, July 8, 1999.

Agence France-Presse. "Economic Crisis Bites Indonesian Python Exporters," *China Daily*, October 28, 2008.

Altieri, Andrew H., Bregje K. van Wesenbeeck, Mark D. Bertness, and Brian R. Silliman. "Facilitation Cascade Drives Positive Relationship Between Native Biodiversity and Invasion Success," *Ecology* 91, no. 5 (2010): 1269-1275.

America's Intelligence Wire. "Officials: Escaped Pet Python Strangled Florida Child," Gale General OneFile, July 1, 2009. http://find.galegroup.com/gtx/start.do?prodld=ITOF.

———. "Villagers Bludgeon Killer Python to Death," Gale General OneFile, November 21, 2003. http://find.galegroup.com/gtx/start.do?prodld=ITOF.

An Act to Establish a National Park Service, and for Other Purposes, 16 U.S.C. § 1 (1916).

An Act to Provide for the Establishment of the Everglades National Park in the State of Florida and for Other Purposes, 16 U.S.C. § 410 (1934).

Anderson, Robert S. "The Lacey Act: America's Premier Weapon in the Fight Against Unlawful Wildlife Trafficking," *Public Land Law Review* (1995).

Animal Welfare Institute, et al. "Joint Statement on H.R. 2811," National Environmental Coalition on Invasive Species, October 30, 2009, http://www.necis.net/files/joint-statement-on-h.r.-2811.

Arave, Lynn. "Utah's Pet Burmese Python Dies at Age 43," *Deseret News*, December 3, 2009.

Architect of the Capitol. Liberty and the Eagle, 2011, http://www.capitol.gov/.

Arnold, John, and David Lyons. "Python Comes Crawling at Vizcaya," *The Miami Herald*, November 27, 1983.

Asbury, John. "Python Dubbed the 'Monster of Lake Evans' Caught in Riverside's

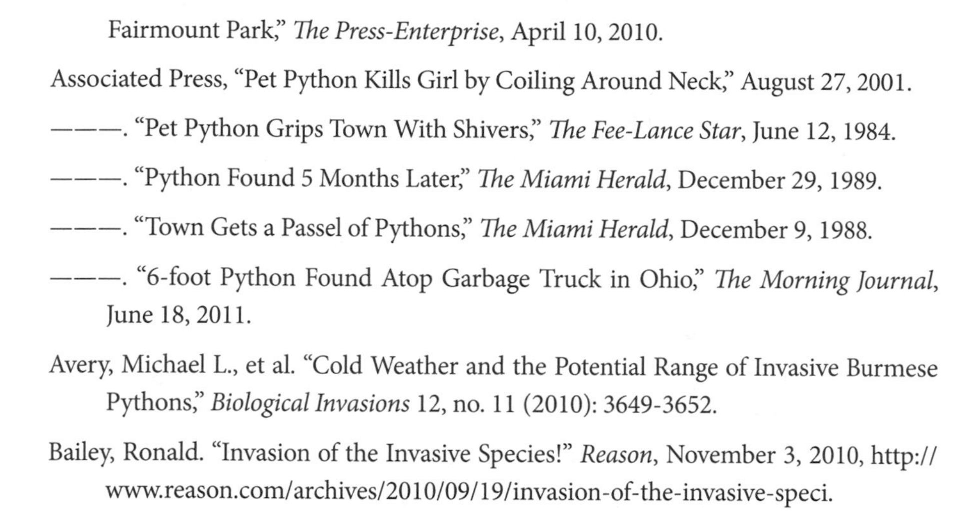

Fairmount Park," *The Press-Enterprise*, April 10, 2010.

Associated Press, "Pet Python Kills Girl by Coiling Around Neck," August 27, 2001.

———. "Pet Python Grips Town With Shivers," *The Fee-Lance Star*, June 12, 1984.

———. "Python Found 5 Months Later," *The Miami Herald*, December 29, 1989.

———. "Town Gets a Passel of Pythons," *The Miami Herald*, December 9, 1988.

———. "6-foot Python Found Atop Garbage Truck in Ohio," *The Morning Journal*, June 18, 2011.

Avery, Michael L., et al. "Cold Weather and the Potential Range of Invasive Burmese Pythons," *Biological Invasions* 12, no. 11 (2010): 3649-3652.

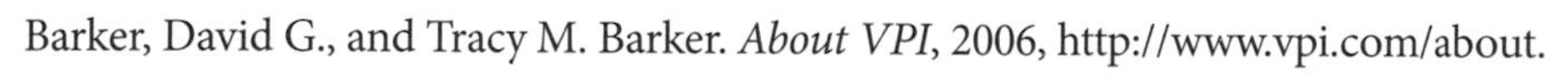

Bailey, Ronald. "Invasion of the Invasive Species!" *Reason*, November 3, 2010, http://www.reason.com/archives/2010/09/19/invasion-of-the-invasive-speci.

Baker, Rochelle. "Reptile Guy Shedding Business," *The Abbotsford Times*, February 18, 2011.

Barker, David G., and Tracy M. Barker. *About VPI*, 2006, http://www.vpi.com/about.

———. *An Open Letter to Dr. Ray W. Snow*, Vida Preciosa International, Inc., 2008.

———. *Climate-matching Predictions for Spread of Giant Snakes in U.S. "Grossly Exaggerated,"* March 13, 2010, http://blogs.nationalgeographic.com/blogs/news/chiefeditor/2010/03/climate-matching-predictions-giant-snakes-exagerrated.html.

———. "Comments on a Flawed Herpetological Paper and an Improper and Damaging News Release from a Government Agency," *Bulletin of the Chicago Herpetological Society* 43, no. 3 (2008): 45-47.

———. "Herpetoculture in the 21st Century," *Bulletin of the Chicago Herpetological Society* 45, no. 9 (2010): 144-149.

———. "On Burmese Pythons in the Everglades," *The Occasional Papers of Vida Preciosa International (VPI International)*, July 2009.

———. "The Distribution of the Burmese Python, *Python molurus bivittatus*," *Bulletin of the Chicago Herpetological Society* 43, no. 3 (2008): 33-38.

———. "The Precautionary Principle and Pythons." *The Occasional Papers of Vida Preciosa International (Vida Preciosa International)*, no. 2 (January 2010).

Barker, David G., James B. Murphy, and Kenneth W. Smith. "Social Behavior in a Captive Group of Indian Pythons, *Python molurus* (Serpentes, Boidae)," *Copeia* 1979, no. 3 (August 1979): 466-471.

Barron, Michael. "Python vs. Gator," *National Geographic Explorer*, National Geographic. August 16, 2006.

Bartelt, Paul E., et al. "Letter to the Subcommittee on Crime, Terrorism & Homeland Security, U.S. House of Representatives Committee on the Judiciary," January 20, 2010.

Bartlett, Richard D., and Patricia P. Bartlett. *A Field Guide to Florida Reptiles and Amphibians*. Houston, TX: Gulf Publishing Company, 1999.

Batchelor, Laura. "Bronx Zoo Cobra That Went Missing is Named Mia," *CNN*, April 7, 2011. http://articles.cnn.com/2011-04-07/us/new.york.cobra.name_1_public-vote-jim-breheny-mia?_s=PM:US.

Beck, Margery A. "Police Say Pet Snake Strangles Suburban Omaha Man," *Fremont Tribune*, June 10, 2010.

Behnke, Patricia. "FWC Orders Amnesty Program for Reptiles of Concern," December 10, 2009, http://myfwc.com/news/news-releases/2009/december/10/rocamnesty/.

———. "Large Python Seized from Crestview Resident's Home," October 23, 2009. http://myfwc.com/news/news-releases/2009/october/23/news_09_nw_crestviewpython/.

Bennetts, Robert E., and Wiley M. Kitchens. "Influence of the 1994-1995 High Water Event on Survival, Reproduction, and Distribution of Snail Kites in the Florida Everglades." In *Ecological Assessment of the 1994-1995 High Water Conditions in the Southern Everglades*, edited by Thomas V. Armentano, 197-204. Miami, FL: South Florida Natural Resources Center, Everglades National Park, 1996.

Bhupathy, S., and V. S. Vijayan. "Status, Distribution and General Ecology of the Indian Python, *Python molurus molurus* Linn., in Keoladeo National Park, Bharatpur, Rajasthan," *Journal of the Bombay Natural History Society* 86, no. 3 (1989): 381-387.

Bilger, Burkhard. "Swamp Things: Florida's Uninvited Predators," *The New Yorker*, April 20, 2009: 80–89.

Black, Richard. "Branson Lemur Plans Bear Fruit," *BBC*, May 23, 2011. http://www.bbc.co.uk/news/science-environment-13502222.

Boone, Christian. "Venomous rattlesnake found dead 100 yards from Zoo Atlanta," *The Atlantic Journal Constitution*, August 30, 2010.

———. "Grant Park Toddler First to Encounter Deadly Snake," *The Atlantic Journal Constitution*, August 31, 2010.

Born Free USA. "Born Free USA Database Sheds Light on Public Safety Issue," news release, February 2, 2011.

U.S. Congress, House. *Nonnative Wildlife Invasion Prevention Act*. HR 669, 111th Cong., 1st Session, January 26, 2009.

Breining, Greg. "Courting Controversy with a New View of Exotic Species," *Yale Environment 360*, November 19, 2009, http://e360.yale.edu/content/feature.msp?id=2212.

———. "The Carp Are Coming! (And Is That Really Such a Bad Thing?)," *Star Tribune*, January 23, 2010.

Bryen, Whitney. "Money Python: Hunters Seek Deadly Snake in Everglades for Kills, Not Dollar Bills," *Naples Daily News*, July 26, 2009.

Burdick, Alan. *Out of Eden: An Odyssey of Ecological Invasion*. New York, NY: Farrar, Straus and Giroux, 2005.

Carey, Clay. "Snake Lovers Fear Transporting Rules Would Hurt Hobby," *The Tennessean*, February 5, 2010.

Carson, Rachel. *The Sense of Wonder*. New York, NY: HarperCollins, 1956.

Cave, Damien. "Pythons in Florida Stalked by Hunters and Tourists Alike," *The New York Times*, May 7, 2010.

Cerabino, Frank. "Lobbyist Hisses: Give Snakes a Fair Shake," *The Palm Beach Post*, August 19, 2009.

Clark, Cammy. "Grassy Key Goal: No Rat Left Behind," *The Miami Herald*, June 8, 2007.

Cocking, Susan. "Pythons 101: Learning to Hunt Everglades Critters," *News & Observer*, March 11, 2010.

———. "Non-native Swamphens Infiltrating Florida's Ecosystem," *The Miami Herald*, May 16, 2010.

Colarossi, Anthony. "Killer-whale Attack: After Human Deaths, Which Animals Die and Which Live?" *Orlando Sentinel*, February 25, 2010.

———. "Killer Python Case: DCF Details Drug Allegations," *The Orlando Sentinel*, August 27, 2009.

Cole, Kevin, and Andrew J. Nelson. "Deadly Snake Attack Called Rare," *Omaha World-Herald*, June 10, 2010.

Collins, Timothy M., and Barbie Freeman. *Genetic Characterization of Populations of the Nonindigenous Burmese Python in Everglades National Park*. Department of Biological Sciences, Florida International University, Miami, FL, 2008.

Cooper, Sam. "Pet Owners Ignore New Law: Few Apply for Permit as Anger Grows Over Rules," *The Province*, April 4, 2010.

Cox, Christian L., and Stephan M. Secor. "Effects of Meal Size, Clutch, and Metabolism on the Energy Efficiencies of Juvenile Burmese Pythons, *Python molurus*," *Comparative Biochemistry and Physiology - Part A: Molecular & Integrative Physiology* 148, no. 4 (December 2007): 861-868.

Dalrymple, George. "Non-Indigenous Amphibians and Reptiles in Florida." In *An Assessment of Invasive Non-Indigenous Species in Florida*, edited by D. C. Schmitz and T. C. Brown, 67-78. Tallahassee, FL: Florida Department of Environmental Protection, 1994.

Daly, Matthew. "Senate Hearing Focuses on Diseases, Invasive Species Threatening People, Native Wildlife," *South Florida Sun Sentinel*, July 8, 2009.

Daugherty, Melissa. "Snake Tales: Proposed Federal Law Threatens Reptile Breeders," *Chico News and Review*, March 4, 2010.

Dauphine, Nico, and Robert J. Cooper. "Pick One: Outdoor Cats or Conservation," *Wildlife Professional*, 2011, Spring ed.

Davis, Mark A. "Researching Invasive Species 50 Years After Elton: A Cautionary Tale," In *Fifty Years of Invasion Ecology: The Legacy of Charles Elton*, by David M. Richardson, 269–275. Blackwell Publishing, Ltd., 2011.

Day, Shawn. "Pet Python Blamed in Virginia Beach Woman's Strangulation," *The Pilot*, 2008.

Defenders of Wildlife, et al. "Joint Statement to the Subcommittee on Insular Affairs, Oceans, and Wildlife on H.R. 669." April 23, 2009.

DogsBite.org. "2009 U.S. Dog Bite Fatalities," *DogsBite.org*, September 21, 2010, http://www.dogsbite.org/bite-fatalities-2009.htm.

Dolinski, Catherine. "Florida Lawmakers Could Force Ban of Burmese Pythons," *The Tampa Tribune*, August 10, 2009.

Donnelly, John. "Exotic: Runaway Creatures Making Extra Work for Trappers, Worry for Veterinarians," *The Miami Herald*, March 28, 1993.

Dorcas, Michael E., John D. Willson, and J. Whitfield Gibbons. "Can Invasive Burmese Pythons Inhabit Temperate Regions of the Southeastern United States?" *Biological Invasions* 13, no. 4 (2010): 793–802.

Dorell, Oren. "Plan to Block Giant Snakes in Florida May Hurt Business," *USA Today*, May 4, 2010.

Dove, Carla J., Ray W. Snow, Michael R. Rochford, and Frank J. Mazzotti. "Birds Consumed by the Invasive Burmese Python (*Python molurus bivittatus*) in Everglades National Park, Florida, USA," *The Wilson Journal of Ornithology*, 2011: 126-131.

Elton, Charles S. *The Ecology of Invasions by Plants and Animals*. London: Redwood Press Limited, 1958.

Enge, Kevin M., Kenneth L. Krysko, Kraig R. Hankins, Todd S. Campbell, and F. Wayne King. "Status of the Nile Monitor (*Varanus niloticus*) in Southwestern Florida," *Southeastern Naturalist* 3, no. 4 (2004): 571–582.

Engeman, Richard M., B. Constantine, and J. Bunting. "The Political, Economic and Management History of a Successful Exotic Eradication: The Case of Black-tailed Jackrabbits in One Part of Florida." In *Managing Vertebrate Invasive Species: Proceedings of an International Symposium*, edited by Witmer, G. W., W. C. Pitt, and K. A. Fagerstone, 476–478. Fort Collins, CO: United States Department of Agriculture, 2007.

Engeman, Richard M., Danny V. Rodriguez, Michael A. Linnell, and Mikel E. Pitzler." A Review of the Case Histories of the Brown Tree Snakes (*Boiga irregularis*) Located by Detector Dogs on Guam," *International Biodeterioration & Biodegradation* 42 (1998): 161–165.

Engeman, Richard M., et al. "Rapid Assessment for a New Invasive Species Threat: the Case of the Gambian Giant Pouched Rat in Florida," *Wildlife Research* 33 (2006): 439–448.

Engeman, Richard M., et al. "The Economic Impacts to Commercial Farms from Invasive Monkeys in Puerto Rico," *Crop Protection* 29 (2010): 401-405.

Engeman, Richard M., et al. "The Path to Eradication of the Gambian Giant Pouched Rat in Florida." In *Managing Vertebrate Invasive Species: Proceedings of an International Symposium*, edited by Witmer, G. W., W. C. Pitt, and K. A. Fagerstone, 305-311. Fort Collins, CO: United States Department of Agriculture, 2007.

Engstrom, Tim. "Southwest Florida Reptile Sales to Slither Away," *Fort Meyers News Press*, March 23, 2010.

Environmental Law Institute. *Status and Trends in State Invasive Species Policy: 2002-2009*. Washington, D.C.: Environmental Law Institute, 2010.

Everest, John W., James H. Miller, Donald M. Ball, and Mike Patterson. *Kudzu in*

Alabama: History, Uses, and Control. Alabama Cooperative Extension System, 1999.

Everglades National Park. Wildlife Observation File, 1979.

Fàbregas, María C., Federico Guillén-Salazar, and Carlos Garcés-Narro. "The Risk of Zoological Parks as Potential Pathways for the Introduction of Non-indigenous Species," *Biological Invasions* 12, no. 10 (2010): 3627–3636.

Faiola, Anthony. "A Slithering Colossus Reigns in Key Biscayne," *The Miami Herald*, March 2, 1991: 1D.

Ferraro, Gabriella. "FWC Busts Miami Man for Selling Pythons on Craigslist Without Proper Licenses." Florida Fish and Wildlife Conservation Commission, May 26, 2010, http://www.fwc.state.fl.us/news/news-releases/2010/may/26/news_10_s_pythonbustmiami/.

Ferraro, Gabriella, and Barron Moody, interview by Missy Tancredi and Jim Jackson. *Radio Green Earth* (December 12, 2009).

Fiala, Jennifer. "Practitioner Revisits Encounter with Monkeypox: Outbreak Lives on for Those Who Experienced the First Known Cases of Monkeypox in the United States," *DVM Newsmagazine*, July 1, 2005.

Flemming, Paul. "Florida Lawmakers Work Proposal to Ban Ownership Exotic Reptiles," *Tallahassee Democrat*, March 2, 2010.

Fleshler, David. "Media Event to Start South Florida Python Hunt Bags 9-footer," *The Sun Sentinel*, July 17, 2009.

———. "Burmese Python Escapes from West Palm Beach House," *The Sun Sentinel*, June 17, 2011.

———. "Everglades Hunt for Burmese Pythons Fails to Catch Any Snakes," *The Sun Sentinel*, April 19, 2010.

Fleshler, David, and Dana Williams. "Wildlife Trade Brings Tarantulas, Pythons, Cobras," *The Sun Sentinel*, November, 30, 2010.

Fleshler, David, and Lisa J. Huriash. "Cold Snap Killed Many Pythons in the Everglades," *The Sun Sentinel*, February 10, 2010.

Florida Fish and Wildlife Conservation Commission. "Captive Wildlife Licenses & Permits," Florida Fish and Wildlife Conservation Commission, July 17, 2011, http://myfwc.com/license/wildlife/captive-wildlife/#not required.

———. "Cold-Related Mortality Event Winter 2009-2010," Florida Fish and Wildlife Conservation Commission, 2010, http://myfwc.com/research/manatee/rescue-mortality-response/mortality-statistics/cold-related-2009-2010/.

———. "Python Removal Program Phase I Summary," Florida Fish and Wildlife Conservation Commission, 2009.

———. "Trapper Who Staged Python Capture Faces Multiple Charges," Florida Fish and Wildlife Conservation Commission, November 9, 2009, http://www.fwc.state.fl.us/NEWSROOM/?p=98&.

———. "Wildlife as Personal Pets," Florida Administrative Code, January 1, 2008.

Fowler, A. J., Lodge, D. M., Hsia, J. F. "Failure of the Lacey Act to Protect US Ecosystems Against Animal Invasions," *Frontiers in Ecology and the Environment* 5, no. 7 (September 2007): 353–359.

Frazer, Gary. "Testimony of Gary Frazer, Assistant Director for Fisheries and Habitat Conservation, U.S. Fish and Wildlife Service, Department of the Interior, before the House Natural Resources Subcommittee on Insular Affairs, Oceans, and Wildlife regarding H.R. 669," April 23, 2009.

French, Thomas. *Zoo Story*. New York, NY: Hyperion, 2010.

Fritts, Ellen I. *Wildlife and People at Risk: A Plan to Keep Rats Out of Alaska*. Alaska Department of Fish and Game, 2007.

Fritts, Thomas H., and David Chiszar. "Snakes on Electrical Transmission Lines: Patterns, Causes, and Strategies for Reducing Electrical Outages Due to Snakes." In *Problem Snake Management: The Habu and the Brown Treesnake*, edited by Gordon H. Rodda, Yoshio Sawai, David Chiszar and Hiroshi Tanaka, 89–103. Ithaca, NY: Cornell University Press, 1999.

Fritts, Thomas H., and Gordon H. Rodda. "The Role of Introduced Species in the Degradation of Island Ecosystems: A Case History of Guam," *Annual Review of Ecology and Systematics* 29 (1998): 113–140.

Fritts, Thomas H., and Michael J. McCoid. "The Threat to Humans from Snakebite by Snakes of the Genus *Boiga* Based on Data From Guam and Other Areas." In *Problem Snake Management: The Habu and the Brown Treesnake*, edited by Gordon H. Rodda, Yoshio Sawai, David Chiszar and Hiroshi Tanaka, 116–127. Ithaca, NY: Cornell University Press, 1999.

Fritts, Thomas H., Michael J. McCoid, and Douglas M. Gomez. "Dispersal of Snakes to Extralimital Islands: Incidents of the Brown Treesnake (*Boiga irregularis*) Dispersing to Islands in Ships and Aircraft." In *Problem Snake Management: The Habu and the Brown Treesnake*, edited by Gordon H. Rodda, Yoshio Sawai, David Chiszar and Hiroshi Tanaka, 209–223. Ithaca, NY: Cornell University Press, 1999.

Gauen, Pat. "Trial for Parents of Little Boy Was Full of Surprises," *St. Louis Post-Dispatch*. April 3, 2000.

Gibson, William. "Snake Lovers Oppose Ban on Burmese Pythons," *The Sun Sentinel*, April 21, 2010.

Gilardi, James D. "Letter to the U.S. House of Representatives Committee on Natural Resources Subcommittee on Insular Affairs, Oceans, and Wildlife from the World Parrot Trust USA, Inc," April 22, 2009.

Gillette, Christopher R., et al. "Oustalet's Chameleon, *Furcifer oustaleti* (Mocquard 1894), a Non-indigenous Species Newly Established in Florida," *IRCF Reptiles & Amphibians* 17, no. 4 (September 2010): 248–249.

Gillingham, J. C. & Chambers, J. A. "Courtship and Pelvic Spur Use in the Burmese Python, *Python molurus bivittatus*," *Copeia* 1982, no. 1 (February 1982): 193–196.

Global Invasive Species Programme. *Best Practices in Pre-Import Risk Screening for Species of Live Animals in International Trade*, edited by Sarah Simons and Maj De Poorter. University of Notre Dame, IN: Global Invasive Species Programme, 2008.

Goode, Stephen. "Hard to Swallow," *Insight on the News*, November 12, 2001.

Goodrich, Robert. "Couple Had No Way To Know Snake Would Kill, Expert Says - Python Breeder Testifies in Trial Over Boy's Death," *St. Louis Post-Dispatch*. March 24, 2000.———. "Couple Whose Son Was Killed by Pet Snake Believes a Jury Will Acquit Them - Centralia, Ill., Parents Have Received Some Hate Mail, Hostile Calls - Trial is Set For Nov. 15," *St. Louis Post-Dispatch*, September 19, 1999.

———. " 'Parents Knew Snake That Killed Child Was Dangerous,' Prosecutor Charges – 'Death Was an 'Unforeseeable and Horrific Tragedy,' Defense Says." *St. Louis Post-Dispatch*, March 22, 2000.

Goodwin, Thomas M., and Wayne R. Marion. "Seasonal Activity Ranges and Habitat Preference of Adult Alligators in a North-Central Florida Lake," *Journal of Herpetology* 13, no. 2 (April 1979): 157–163.

Gorman, Bill. "SyFy Saturday Original Movie 'Mega Python vs Gatoroid' Devours 2.35 Million Total Viewers," *TV by the Numbers*, February 1, 2011, http://tvbythenumbers.zap2it.com/2011/02/01/syfy-saturday-original-movie-mega-python-vs-gatoroid-devours-2-35-million-total-viewers/81125.

Gorman, James. "A Diet for an Invaded Planet: Invasive Species," *The New York Times*, January 2, 2011.

Grace, Michael S., Owen M. Woodward, Don R. Church, and Gwen Calisch. "Prey Targeting by the Infrared-imaging Snake *Python molurus*: Effects of Experimental and Congenital Visual Deprivation," *Behavioural Brain Research* 119, no. 1 (February 2001): 23–31.

Gray, Denis D. "14-foot Python Attacks Handler." America's Intelligence Wire, December 31, 2006.

Graziani, Greg, Shawn Heflick, and Michael Cole. "Facts about Burmese Pythons and Other Reptiles of Concern," *Reptile Clan Rescue*, December 25, 2009, http://www.reptileclan.com/news.php?id=9.

Green, Alan. *Animal Underworld: Inside America's Black Market for Rare and Endangered Species*. New York, NY: Public Affairs, 1999.

Griffiths, Christine J., Dennis M. Hansen, Carl G. Jones, Nicolas Zuel, and Stephen Harris. "Resurrecting Extinct Interactions with Extant Substitutes," *Current Biology* 21 (2011): 1–4.

Grimes, Brian K. *NRC Information Notice 93–53: Effect of Hurricane Andrew on Turkey Point Nuclear Generating Station and Lessons Learned*. NRC Washington, D.C.: United States Nuclear Regulatory Commission, 1993.

Grimm, Fred. "Pythons Need Culinary PR, Not Bounties," *The Miami Herald*, June 2, 2009.

Groot, T., E. Bruins, and J. Breeuwer. "Molecular Genetic Evidence for Parthenogenesis in the Burmese Python, *Python molurus bivittatus*," *Heredity* 90 (2003): 130–135.

Haas, Brian. "Piranha Problem," *The Sun Sentinel*, November 18, 2009.

Hambleton, Laura, and John Donnelly. "Releasing Unwanted Exotic Animals is Harmful, Illegal," *The Miami Herald*, August 31, 1991.

Hancock, David. "Snake Stops Bulldozer in Its Tracks," *The Miami Herald*, October 7, 1989.

Hansen, Dennis M., C. Josh Donlan, Christine J. Griffith, and Karl J. Campbell. "Ecological History and Latent Conservation Potential: Large and Giant Tortoises as a Model for Taxon Substitutes," *Ecography* 33 (2010): 272–284.

Hardin, Scott. "Testimony of Scott Hardin, Exotic Species Coordinator, Florida Fish and Wildlife Conservation Commission, Joint Oversight Hearing of the Subcommittee, March 23, 2010." Washington, D.C., March 23, 2010.

Harrison, David. "Sir Richard Branson's 'Eco-island' Plans Hit by Row Over Non-native Lemurs," *The Telegraph*, April 16, 2011.

Harvey, Rebecca G., et al. *Burmese Pythons in South Florida: Scientific Support for Invasive Species Management*. Davie, FL: University of Florida Institute of Food and Agricultural Sciences, 2008.

Haseltine, Susan. "USGS Defends Study That Suggests U. S. Climate May Become Accommodating to Giant Alien Snakes," *National Geographic*, January 23, 2010, http://blogs.nationalgeographic.com/blogs/news/chiefeditor/2010/01/usgs-defends-study-thast-sugge.html.

Hecker, Charles E. "Mammoth Serpent Surfaces in Davie 18-foot, 200-pound Python is Dug from Shed," *The Miami Herald*, January 27, 1991.

Hedin, Jonas, and Thomas Ranius. "Using Radio Telemetry to Study Dispersal of the Beetle *Osmoderma eremita*, an Inhabitant of Tree Hollows," *Computers and Electronics in Agriculture* 35, no. 2–3 (August 2002): 171–180.

Hench, David. "Abandoned Viper in Saco Puzzles Aficionados," *The Portland Press Herald*, March 13, 2010.

Herald Tribune Staff. "Reptiles of Big Concern," *The Sarasota Herald Tribune*, October 17, 2009.

Herszenhorn, David. "13-foot-long Pet Python Kills Its Caretaker," *The New York Times*, October 10, 1996.

Hiaasen, Rob. "Reality Bites: Inside an Indonesian Cave, Severna Park's Brady Barr, a TV Show Host, Meets a Python With a Good Grip," *The Baltimore Sun*, October 9, 2007.

Hudak, Stephen. "Jurors Find Couple Guilty of all Charges in Killer-python Case," *The Orlando Sentinel*, July 14, 2011.

———. "Pet Python Likely Starving When It Killed Sumter County Child, Documents Show," *The Orlando Sentinel*, January 1, 2011.

Hughes, Colin. *An Assessment of the Threat Posed to the Everglades National Park Ecosystem by Potentially Invasive Tegu Lizards*. Department of Biological Sciences, Florida Atlantic University, Davie, FL, 2010.

Hughes, Sallie. "Escaped Snake Had Mice Time," *The Miami Herald*, September 28, 1985.

Hunt, Jared. "Missing Python Drama Inspires Exotic Animal Ordinance," *Charleston Daily Mail*, June 17, 2010.

———. "Town Will Require Owners of Exotic Pets to Get License," *Charleston Daily Mail*, July 8, 2010.

Inambao, Chrispin, "Python Victim Uncoils His Tale," *Africa News Service*, February 27, 2001.

IPCC. *IPCC Fourth Assessment Report (AR4)*. Geneva, Switzerland: Intergovernmental

Panel on Climate Change, 2007.

Jacobs, H. J., M. Auliya, and W. Bohme. "On the Taxonomy of the Burmese Python, Python molurus bivittatus, Kuhl, 1820, Specifically on the Sulawesi Population." Sauria 31, no. 3 (2009): 5-16.

Jacobson, Elliott, et al. "Letter to the Subcommittee on Crime, Terrorism & Homeland Security, U.S. House of Representatives Committee on the Judiciary." November 24, 2009.

Jeffrey, Nancy. "Everglades Campers Kill Boa Constrictor," *The Miami Herald*, September 3, 1989.

Jenkins, P. T., K. Genovese, and H. Ruffler. *Broken Screens: The Regulation of Live Animal Imports in the United States*. Washington, D.C.: Defenders of Wildlife, 2007.

Jinbo, Paige L. "Snake's Alive! A Boa Constrictor Discovered in Waiawa Gulch Was Probably Someone's Pet," *Honolulu Star-Advertiser*, July 6, 2011.

Johnson, O'Ryan. "Path to Recovery," *The Boston Herald*, June 11, 2011.

Johnson, Steve, and Monica McGarrity. *Florida's Introduced Birds: Sacred Ibis (Threskiornis aethiopicus)*. University of Florida IFAS Extension, 2009, http://edis.ifas.ufl.edu/uw312.

Jones, George Pierce. *The Feasibility of Using Small Unmanned Aerial Vehicles for Wildlife Research*. Gainesville, FL: University of Florida, 2003.

Jones, Rachel. "7 Trappers Coax 20-foot Python from Under Home," *The Miami Herald*, August 18, 1989.

Kam, Dara. "Game On: Crist Orders Python Purge," *The Palm Beach Post*, July 15, 2009.

Kaufman, Leslie. "Snake Owners See Furry Bias in Invasive Species Proposal," *The New York Times*, January 8, 2011.

Kessler, Rebecca. "Mercury Keeps Invasive Pythons Off the Menu," *LiveScience*, August 31, 2010.

KHOU. "Albino Python Kills Chicken, Goat, Dog in Texas," *KHOU*, July 15, 2011, http://www.khou.com.

Kiley, Kevin. "N.C. Clamps Down on Deadly, Exotic Snakes," *The Charlotte Observer*, July 10, 2009.

King, W., and T. Krakauer. "The Exotic Herpetofauna of Southeast Florida," *Quarterly*

Journal of the Florida Academy of the Sciences 29 (1966): 144–154.

Kluger, Jeffrey. "Killer-Whale Tragedy: What Made Tilikum Snap?" *Time*, February 26, 2010.

Kolar, Cynthia S., and David M. Lodge. "Progress in Invasion Biology: Predicting Invaders," *TRENDS in Ecology & Evolution* 16, no. 4 (April 2001): 199–204.

Koss, Geof. "Snake Lobby, Animal Groups Clash Over Python Bills," *Roll Call*, August 16, 2009.

Krysko, Kenneth L., et al. "The African Five-Lined Skink, *Trachylepis quinquetaeniata* (Lichtenstein 1823): A Newly Established Species in Florida," *IRCF Reptiles & Amphibians* 17, no. 3 (September 2010): 183–184.

Krysko, Kenneth L, et al. "Verified non-indigenous amphibians and reptiles in Florida from 1863 through 2010: Outlining the invasion process and identifying invasion pathways and stages," *Zootaxa* 3028 (2011): 1-64.

Krysko, Kenneth L., James C. Nifong, Ray W. Snow, Kevin M. Enge, and Frank J. Mazzotti. "Reproduction of the Burmese Python (*Python molurus bivittatus*) in Southern Florida," *Applied Herpetology* 5 (2008): 93–95.

Labisky, Ronald F., Karl E. Miller, and Christine S. Hartless. "Effect of Hurricane Andrew on Survival and Movements of White-tailed Deer in the Everglades," *The Journal of Wildlife Management* 6, no. 3 (July 1999): 872–879.

Langley, Ricky L. "Human Fatalities Resulting From Dog Attacks in the United States, 1979–2005," *Wilderness and Environmental Medicine* 20 (2009): 19–25.

Lee, Julian C. "*Anolis sagrei* in Florida: Phenetics of a Colonizing Species I. Meristic Characters," *Copeia*, no. 1 (1985): 182–194.

Leopold, Aldo. *A Sand County Almanac*. New York: Oxford University Press, 1949.

Lodge, David M. "Testimony by David M. Lodge Before the House Natural Resources Subcommittee on Insular Affairs, Oceans, and Wildlife." April 23, 2009.

Lodge, Thomas E. *The Everglades Handbook*. 3rd Edition. Boca Raton, FL: CRC Press, 2010.

Louv, Richard. *Last Child in the Woods: Saving Our Children from Nature-Deficit Disorder*. Chapel Hill, NC: Algonquin Books of Chapel Hill, 2008.

Lowe, S., S. Boudjelas, and M. De Poorter. *100 of the World's Worst Invasive Alien Species: A Selection from the Global Invasive Species Database*. Auckland, New Zealand: Invasive Species Specialist Group, 2000.

Lystra, Tony. "Eight-foot Anaconda Seized from Longview Home," *The Daily News*,

July 14, 2011.

Maehr, David S., E. Darrell Land, David B. Shindle, Oron L. Bass, and Thomas S. Hoctor. "Florida Panther Dispersal and Conservation," *Biological Conservation* 106 (2002): 187–197.

Mann, Charles. *1491: New Revelations of the Americas Before Columbus*. New York, NY: Vintage Books, 2006.

Markowitz, Arnold. "Python Pulled from Bay," *The Miami Herald*, November 13, 1992: 3B.

Martin Jr., William R. "Written Testimony for HR 669 'Nonnative Wildlife Invasion Prevention Act.'" April 23, 2009.

Martinez, Chris. "Man Walking Dogs Comes Face to Face with Python," *ABC Action News*, January 28, 2011, http://www.abcactionnews.com/dpp/news/region_north_pinellas/tarpon_springs/man-walking-dogs-come-face-to-face-with-python.

Martinez-Morales, Miguel Angel, and Alfredo D. Cuaron. "*Boa constrictor*, an Introduced Predator Threatening the Endemic Fauna on Cozumel Island, Mexico," *Biodiversity and Conservation* 8 (1999): 957–963.

Maslow, A. H. "A Theory of Human Motivation," *Psychological Review* 50 (1943): 370–396.

Mazzotti, Frank J., et al. "Cold-induced mortality of invasive Burmese pythons in south Florida," *Biological Invasions* 13, no. 1 (June 2010): 143–151.

Mazzotti, Frank J., Matthew L. Brien, Michael S. Cherkiss, and Skip Snow. *Removing Burmese Pythons from Lands Managed by the South Florida Water Management District*. Department of Wildlife Ecology and Conservation, Fort Lauderdale Research and Education Center, Fort Lauderdale, FL: University of Florida, 2007.

McClellan, Bill. "For Couple, The Trial Ends But Not The Nightmare," *St. Louis Post-Dispatch*, March 26, 2000.

Medzerian, David. "Slithery Character Can't Elude Police," *The Miami Herald*, December 7, 1988.

Meshaka, Jr., Walter E. "A Runaway Train in the Making: The Exotic Amphibians, Reptiles, Turtles, and Crocodilians of Florida. Monograph 1," *Herpetological Conservation and Biology* 6 (2011): 1–101.

———. *The Cuban Treefrog in Florida: Life History of a Successful Colonizing Species*. Gainesville, FL: University Press of Florida, 2001.

Meshaka, Jr., Walter E., Brian P. Butterfield, and Brian J. Hauge. *The Exotic Amphibians and Reptiles of Florida*. Malabar, FL: Krieger Publishing Company, 2004.

Meshaka, Jr., Walter E., William F. Loftus, and Todd Steiner. "The Herpetofauna of Everglades National Park," *Florida Scientist* 63, no. 2 (2000): 84–103.

Meyers, Marshall. "Testimony of Marshall Meyers, Pet Industry Joint Advisory Council, before the Subcommittee on Insular Affairs, Oceans, and Wildlife, House Natural Resources Subcommittee." April 23, 2009.

Miami Herald Staff. "Big Python on the Loose in Pompano," *The Miami Herald*, June 12, 1992.

———. "Two Pythons Slither Their Way to Freedom," *The Miami Herald*, September 11, 1982.

Morelli, Keith. "200-pound Snake Killed in Okeechobee Adds to Mounting Python Woes," *The Tampa Tribune*, July 31, 2009.

———. "Officials: Trapper Faked Capture of 14-foot Python in Bradenton," *The Tampa Tribune*, August 31, 2009.

———. "State commission approves ban on large exotic reptiles," *The Tampa Tribune*, April 28, 2010.

———. "Wildlife Experts Question Python Numbers in Everglades," *The Tampa Tribune*, September 4, 2009.

Morgan, Curtis. "Burmese python tops list of invasive killers," *The Miami Herald*, July 8, 2009.

———. "Exotic Invasion: Pythons Back in the Everglades," *The Miami Herald*, February 7, 2011.

———. "Florida's Pythons Could Have a Bounty on Their Head—and Body," *The Miami Herald*, May 28, 2009.

———. "It's Alien Versus Predator in Glades Creature Clash," *The Miami Herald*, October 5, 2005.

———. "Python Hunter: Miami Man Takes on Glades Swamp Serpents," *The Miami Herald*, August 8, 2009.

———. "Scientists Brace for Snake Invasion," *The Miami Herald*, October 17, 2005.

———. "The Main Event: Gator vs. Python," *The Miami Herald*, January 26, 2003.

———. "The Python Posse Is Coming to the Everglades," *The Miami Herald*, July 14, 2009.

National Weather Service. "Summary of Historic Cold Episode of January 2010," National Oceanic and Atmospheric Administration, Miami, Florida, 2010.

Natural England. *Horizon Scanning for New Invasive Non-native Animal Species in England.* Sheffield, UK: Natural England, 2009.

Nolin, Robert. "Nasty Nile Monitors Showing Up in South Florida," The Sun Sentinel, July 5, 2011.

Noticias EFE. "La Culebra Real, que Llegó a Canarias como Mascota, Amenaza la Fauna Local," *El Norte de Castilla*, June 16, 2011.

Nunez, Martin A., and Daniel Simberloff. "Invasive Species and the Cultural Keystone Concept," *Ecology and Society* 10, no. 1 (2005): r4.

O'Conner, Terry. "Python Invasion Slithering to SW Fla," *Cape Coral Daily Breeze*, October 10, 2009.

O'Connor, Anahad. "Woman Mauled by Chimp Has Surgery, and Her Vital Signs Improve," *The New York Times*, February 18, 2009.

Parker, Molly. "Constrictors Seen as Threat in State," *The Clarion Ledger*, October 26, 2010.

Pearson, Matthew. "Permits Needed for Exotic Pets: New B.C. Law Brings Order to Wild Kingdom," *Victoria Times Colonist*, November 16, 2009.

Perry, Neil D., et al. "New Invasive Species in Southern Florida: Gambian Rat (*Cricetomys gambianus*)," *Journal of Mammology* 87, no.2 (2006): 427–432.

Pet Industry Joint Advisory Council. "About PIJAC," 2010, http://www.pijac.org/about/.

———. "Florida's Senator Nelson Proposes Congressional Ban of All Pythons Under the Lacey Act," February 17, 2009, http://www.pijac.org/_documents/us_sb_373.pdf.

Phillips, Jim. "Killer Bear's Owner Brought a Bear to Athens in 1990," *The Athens News*, August 26, 2010.

Pimental, D., L. Lach, R. Zuniga, and D. Morrison. "Environmental and Economic Costs of Nonindigenous Species in the United States," *BioScience* 50, no. 1 (2000): 53–65.

Pittman, Craig. "Toxic Metal Laces Snakes," *The St. Petersburg Times*, September 6, 2009.

Plunkett, Judy. "Feline Up a Tree is Rescued from Hungry Python," *The Miami Herald*, December 4, 1993.

Poucher, C. "Eradication of the Giant African Snail in Florida." In *Proceedings of the Florida State Horticultural Society*, 523–524. Lake Alfred, FL: Florida State Horticultural Society, 1975.

Power, Trish. "Wanted: One Missing Python Friendly Floydie, A 'Gentle' Snake, Slips Out of Home," *The Miami Herald*, June 19, 1990.

Pyron, A. R., Burbrink, F. T., Guiher, T. J. "Claims of Potential Expansion throughout the U.S. by Invasive Python Species Are Contradicted by Ecological Niche Models," *PLoS ONE* 3, no. 8 (August 2008): e2931.

Quick, John S., Howard K. Reinert , Eric R. De Cuba, and R. Andrew Odum. "Recent Occurrence and Dietary Habits of *Boa constrictor* on Aruba, Dutch West Indies," *Journal of Herpetology* 39, no. 2 (2005): 304–307.

Reed, Robert N., and Gordon H. Rodda. *Giant Constrictors: Biological and Management Profiles and an Established Risk Assessment for Nine Large Species of Pythons, Anacondas, and the Boa Constrictor*. Open-File Report 2009–1202, Reston, Virginia: United States Geological Survey, 2009.

Reed, Robert N., et al. "A Field Test of Attractant Traps for Invasive Burmese Pythons (*Python molurus bivittatus*) in Southern Florida," *Wildlife Research* 38 (2011): 114–121.

Reed, Robert N., et al. "Natural History Notes: *Python sebae*," *Herpetological Review* 42, no. 2 (2011): 303.

Reinert, Howard K., and David Cundall. "*An Improved Surgical Implantation Method for Radio-Tracking Snakes*," *Copeia* 1982, no. 3 (August 1982): 702–705.

Rodda, Gordon H., Catherine S. Jarnevich, and Robert N. Reed. "Challenges in Identifying Sites Climatically Matched to the Native Ranges of Animal Invaders," *PLoS ONE* 6, no. 2 (2011): 1–18.

———. "What Parts of the US Mainland Are Climatically Suitable for Invasive Alien Pythons Spreading from Everglades National Park?" *Biological Invasions* 11, no. 2 (2008): 241–252.

Rodda, Gordon H., Thomas H. Fritts, Michael J. McCoid, and Earl W. Campbell III. "An Overview of the Biology of the Brown Treesnake (*Boiga irregularis*), a Costly Introduced Pest on Pacific Islands." In *Problem Snake Management: The Habu and the Brown Treesnake*, edited by G.H., Sawai, Y., Chiszar, D., Tanaka, H. Rodda, 44-80. Ithaca, NY: Cornell University Press, 1999.

Rodrigues, Jill. "Chimps, Alligators, More Prohibited As Pets," *EastBayRI.com*, March

19, 2010. http://www.eastbayri.com/.

Samuels, Robert. "Critters Capture Capitol Culture," *The St. Petersburg Times*, March 10, 2010.

———. "State Poised to Ban the Sale of Burmese Pythons," *The Miami Herald*, April 29, 2010.

Sarasota Herald-Tribune Staff. "Land of the Reptiles," *The Sarasota Herald Tribune*, August 10, 2009.

Sax, Dov F., and Steven D. Gaines. "Species Diversity: From Global Decreases to Local Increases," *TRENDS in Ecology and Evolution* 18, no. 11 (2003): 561–566.

Schall, Mitchell, interview by David Sutta. "Glades Python Hunt Gets Personal," *WFOR News*, March 8, 2010.

Scheck, Justin. "Bear Market in Boas: Proposed Laws Strangle Sales of Mutant Snakes," *The Wall Street Journal*, February 13, 2010.

Schneider, Amy. "Lionfish Could Upset Ecological Balance in the Caribbean Sea," *The Daily Barometer*, May 5, 2010.

Secor, Stephen M., and Jared Diamond. "Adaptive Responses to Feeding in Burmese Pythons: Pay Before Pumping," *The Journal of Experimental Biology* 198 (1995): 1313–1325.

Secretariat of the Convention on Biological Diversity. "Pets, Aquarium, and Terrarium Species: Best Practices for Addressing Risks to Biodiversity." Technical Series No. 48, United Nations Environment Programme, Montreal, 2010.

Sheerha, Thomas J. "Ohio Bear Kills Caretaker; Owner Had Legal Trouble," *The Boston Globe*, August 20, 2010.

Shwiff, Stephanie A., Karen Gebhardt, Katy N. Kirkpatrick, and Steven S. Shwiff. "Potential Economic Damage from Introduction of Brown Tree Snakes, *Boiga irregularis* (Reptilia: Colubridae), to the Islands of Hawai☒i," *Pacific Science* 64, no. 1 (2010): 1–10.

Smith, T. R., L. A. Whilby, and A. I. Derksen. *2010 Florida CAPS Giant African Snail Survey Report*. Division of Plant Industry, Florida Department of Agriculture and Consumer Services, Gainesville, FL: Florida Department of Agriculture and Consumer Services, 2010.

Snow, Ray W., Kenneth L. Krysko, Kevin M. Enge, Lori Oberhofer, Alice Warren-Bradley, and Laurie Wilkins. "Introduced Populations of *Boa constrictor* (Boidae) and *Python molurus bivittatus* (Pythonidae) in Southern Florida." In *Biology of the Boas and Pythons*, edited by Robert W. & Powell, Robert Henderson, 417–

438. Eagle Mountain, UT: Eagle Mountain Publishing, 2007.

Snow, Ray W., Matthew L. Brien, Michael S. Cherkiss, Laurie Wilkins, and Frank J. Mazzotti. "Dietary Habits of the Burmese python, *Python molurus bivittatus*, in Everglades National Park, Florida," *Herpetological Bulletin*, no. 101 (2007): 5–7.

Snyder, Jim. "Snakes Get Lobbyists in Fight Over Boa Ban," *Bloomberg*, August 5, 2011.

Somaweera, Ruchira, Nilusha Somaweera, and Richard Shine. "Frogs Under Friendly Fire: How Accurately Can the General Public Recognize Invasive Species?" Biological Conservation 143, no. 6 (2010): 1477–1484.

Somma, Louis A. " *Nerodia fasciata pictiventris* (Cope, 1895)," USGS Nonindigenous Aquatic Species Database, October 27, 2009, http://nas.er.usgs.gov/queries/FactSheet.asp?speciesID=1195.

South Florida Water Management District. *2011 South Florida Environmental Report*. West Palm Beach, FL: South Florida Water Management District, 2011.

South Florida Water Management District. "Pythons Persist in Everglades Through Freezes and Water Shortage," news release, March 26, 2011.

Spinner, Kate. "Another species of python is raising concern," *The Sarasota Herald-Tribune*, November 3, 2009.

Spring, Suzanne. "Python Finds Perch in Coral Gables Tree," *The Miami Herald*, March 1, 1983.

Springborn, Michael, Christina M. Romagosa, and Reuben P. Keller. "The Value of Nonindigenous Species Risk Assessment in International Trade," *Ecological Economics* 70, no. 11 (2011): 2145–2153.

Staats, Eric. "Officials Hope Hunters, Public Help Them Root Out, Hunt Down Invading Pythons," *Naples Daily News*, March 7, 2010.

———. "11-foot Python Captured at Rookery Bay," *Naples Daily News*, March 5, 2010.

Steele, Joseph Ryan. "*Tupinambis merianae* Profile Update." Boca Raton, FL: Florida Atlantic University, 2010.

Steinmetz, Katy. "Exotic Animals Non Grata in Ohio," *Time*, January 7, 2011.

Stejneger, L. "Two Geckos New to the Fauna of the United States," Copeia 1922 (1922): 56.

Stevenson, Tommy. "Expert to Study Snakes' Northward Migration," *The Gadsden Times*, August 22, 2009.

Stith, Bradley M., Daniel H. Slone, and James P. Reid. *Review and Synthesis of Manatee Data in Everglades National Park*. United States Department of the Interior,

United States Geological Survey, 2006.

Strickland, Tom. "Let's Figure Out How to Stop Invaders Like the Asian Carp," *Detroit Free Press*, December 7, 2009.

Strong, Allan M., and G. Thomas Bancroft. "Postfledging Dispersal of White-crowned Pigeons: Implications for Conservation of Deciduous Seasonal Forests in the Florida Keys," *Conservation Biology* 8, no. 3 (1994): 770–779.

Sullivan, Dennis. "Orland Park Moves to Outlaw Exotic Pets," *The Chicago Tribune*, October 21, 2010.

Sword, Doug. "Python Population Expected to Explode," *Sarasota Herald-Tribune*, June 1, 2009.

Tampa Tribune Staff. "12-foot Anaconda Found in Kissimmee by Deputies." *The Tampa Tribune*, January 19, 2010.

Taylor, Joe. "9-foot-python Found Under Marco Island Pool Deck," *The News-Press*, March 11, 2010.

Tennant, Alan. *A Field Guide to Snakes of Florida*. Houston, TX: Gulf Publishing Company, 1997.

The Atlanta Journal-Constitution. "Parents Charged in Death," Gale General OneFile, September 2, 1999, http://find.galegroup.com/gtx/start.do?prodld=ITOF.

The Calgary Herald. "Tiger Mauls Woman," May 12, 2007.

The Denver Post. "Owner Killed by Snake Had Been Warned," Gale General OneFile, February 12, 2002, http://find.galegroup.com/gtx/start.do?prodld=ITOF.

———. "Python That Killed Owner Put Down," Gale General OneFile, February 20, 2002, http://find.galegroup.com/gtx/start.do?prodld=ITOF.

The Humane Society of the United States. "The HSUS Calls for Stricter Regulations After Woman Killed by Pet Python," news release, October 24, 2008.

The New York Times. "Python Suspected in Death," Gale General OneFile. July 22, 1993, http://find.galegroup.com/gtx/start.do?prodld=ITOF (accessed July 16, 2009).

The Pittsburg Tribune-Review. "Police Use Taser to Detach Python from Man's Arm," Gale General OneFile, December 2, 2006, http://find.galegroup.com/gtx/start.do?prodld=ITOF.

The Record. "Boy Rescued After Python Sinks Fangs into His Face," Gale General OneFile, April 27, 2001, http://find.galegroup.com/gtx/start.do?prodld=ITOF.

———. "Cop pries python off shop owner," Gale General OneFile, April 20, 2008, http://find.galegroup.com/gtx/start.do?prodld=ITOF.

Tomb, Geoffrey. "Ghostly Hunter Haunts Estate," *The Miami Herald*, January 10, 1995.

Tompkins, Shannon. "Texas Tries to Control Invasion of Exotic Snakes," *The Houston Chronicle*, February 27, 2008.

Trischitta, Linda. "Livestock-killing, 11-foot Burmese Python Nabbed in Miami-Dade," *The Sun Sentinel*, August 1, 2011.

Trousdale, Austin W., and David C. Beckett. "Characteristics of Tree Roosts of Rafinique's Big-Eared Bat (*Corynorhinus rafinesquii*) in Southeastern Mississippi," *American Midland Naturalist* 154, no. 2 (2005): 442–449.

United Press International. "Python Missing for 2 Years is Found," February 2, 2011.

United States Congress Office of Technology Assessment. *Harmful Non-Indigenous Species in the United States*. Washington, D.C.: Government Printing Office, 1993.

United States Fish and Wildlife Service. "Pet Owners Unduly Alarmed Over Proposed Wildlife Import Regulation," news release, August 1, 1974.

———. *South Florida Multi-Species Recovery Plan*. Atlanta, GA: United States Fish and Wildlife Service, 1999.

United States Geological Survey. "History of the Brown Treesnake Invasion on Guam," July 26, 2005, http://www.fort.usgs.gov/resources/education/bts/invasion/history.asp.

United States Government Accountability Office. *Live Animal Imports: Agencies Need Better Collaboration to Reduce the Risk of Animal-related Diseases*. Washington, D.C.: Government Accountability Office, 2010.

———. *South Florida Ecosystem: Restoration Is Moving Forward but Is Facing Significant Delays, Implementation Challenges, and Rising Costs*. Washington, D.C.: United States Government Accountability Office, 2007.

UPI. "Indiana Man Killed by Pet Python," Gale General OneFile, September 5, 2006, http://find.galegroup.com/gtx/start.do?prodld=ITOF.

———. "Pet Python Strangles Ohio Man," Gale General OneFile, December 16, 2006, http://find.galegroup.com/gtx/start.do?prodld=ITOF.

———. "Python Blamed for Death Euthanized," Gale General OneFile, October 28, 2008, http://find.galegroup.com/gtx/start.do?prodld=ITOF.

———. "Python Owner to Pay for Dog's Death," Gale General OneFile, May 10, 2006, http://find.galegroup.com/gtx/start.do?prodld=ITOF.

———. "Fla. Angler Catches 11-foot Burmese Python," November 15, 2005.

Van Mierop, L. H. S., and Susan M Barnard. "Observations on the Reproduction of *Python molurus bivittatus* (Reptilia, Serpentes, Boidae)," *Journal of Herpetology* 10, no. 4 (1976): 333–340.

———. "Observations on Thermoregulation in the Brooding Female *Python molurus bivittatus* (Serpentes: Boides)," *Copeia* 1978, no. 4 (1978): 615–621.

Vernon, Amy. "Recovering Escaped Python May Put Squeeze on Owner," *The Miami Herald*, August 28, 1993.

Wadlow, Kevin. "A Signal Rats Out Keys Python: It's 1st Time Exotic Species Found in Islands, Experts Say," *Florida Keys Keynoter*, April 28, 2007.

Wahlberg, David. "Monkeypox Spurs Ban on Rodent Imports, Sales," *The Atlanta Journal-Constitution*, June 12, 2003.

Wall, Frank. *Ophidia Taprobanica or the Snakes of Ceylon*. Government Printer, Colombo, 1921.

Wikelski, Martin, David Moskowitz, James S. Adelman, Jim Cochran, David S. Wilcove, and Michael L. May. "Simple Rules Guide Dragonfly Migration," *Biology Letters* 2 (2006): 325–329.

Wiles, Gary J., Jonathan Bart, Jr., Robert E. Beck, and Celestino F. Aguon. "Impacts of the Brown Tree Snake: Patterns of Decline and Species Persistence in Guam's Avifauna," *Conservation Biology* 17, no. 5 (2003): 1350–1360.

Willson, J. D., M. E. Dorcas, and R. W. Snow. "Identifying Plausible Scenarios for the Establishment of Invasive Burmese Pythons (*Python molurus*) in Southern Florida," *Biological Invasions* 13 (2010): 1493–1504.

Wilson, D. S., H. R. Mushinsky, and R. A. Fischer. *Species Profile: Gopher Tortoise (Gopherus Polyphemus) on Military Installations in the Southeastern United States*. Washington, D.C.: Strategic Environmental Research and Development Program, 1997.

Wilson, L. D., and L. Porras. *The Ecological Impact of Man on the South Florida Herpetofauna*. Lawrence, KS: University of Kansas / World Wildlife Fund, 1983.

Winchester, Chris. *An Evaluation of Habitat Selection and an Abundance Estimate for the Endangered Key Largo Woodrat*. Masters Thesis, Athens, GA: University of Georgia, 2007.

Woodlee, Yolanda. "Man Comes to the Aid of 200-Pound Python," *The Miami Herald*, October 3, 1986.

Wyatt, Andrew. "Testimony Before the Committee on the Judiciary, Subcommittee on

Crime, Terrorism, and Homeland Security." November 5, 2009.

Yoon, Carol Kaesuk. "Ecological Menace Could Shatter Hawaii's Freedom From Snakes," *The New York Times*, July 14, 1992.

Zoo Miami. "Zoo History," 2006, http://www.miamimetrozoo.com/about-metro-zoo.asp?Id=93&rootId=8.

Index

(References to color insert photos, maps, and charts appear in bold.)

About the Author

Larry Perez is a lifelong resident of Miami who has spent over fifteen years working in south Florida's natural areas. During his career, he has worked as a naturalist for Miami-Dade Parks and Recreation, and as a ranger for Biscayne and Everglades National Parks. Larry is a graduate of Florida International University where he completed programs in park and recreation management and environmental studies. He is also the author of *Words on the Wilderness: A History of Place Names in South Florida's National Parks* (ECity Publishing, 2007) and maintains a healthy fascination with lizards and snakes.

Here are some other books from Pineapple Press on related topics. For a complete catalog, visit our website at www.pineapplepress.com. Or write to Pineapple Press, P.O. Box 3889, Sarasota, Florida 34230-3889, or call (800) 746-3275.

Everglades: River of Grass, 60th Anniversary Edition by Marjory Stoneman Douglas with an update by Michael Grunwald. Before 1947, when Marjory Stoneman Douglas named the Everglades a "river of grass," most people considered the area worthless. She brought the world's attention to the need to preserve the Everglades. In the Afterword, Michael Grunwald tells us what has happened since Douglas rallied Floridians to her cause to save the Glades. (hb)

Florida Magnificent Wilderness: State Lands, Parks, and Natural Areas by James Valentine and D. Bruce Means. Photographer James Valentine captures environmental art images of the state's remote wilderness places. Dr. D. Bruce Means covers the wildlife and natural ecosystems of Florida. An introduction to each section is written by a highly respected Florida writer and conservationist, including Al Burt, Manley Fuller, Steve Gatewood, Victoria Tschinkel, and Bernie Yokel. (hb)

Florida's Birds, 2nd Edition by David S. Maehr and Herbert W. Kale II. Illustrated by Karl Karalus. This new edition is a major event for Florida birders. Each section of the book is updated, and 30 new species are added. Also added are range maps and color-coded guides to show the months when each bird is present and/or breeding in Florida. Color throughout. (pb)

Florida's Rivers by Charles R. Boning. An overview of Florida's waterways and detailed information on 60 of Florida's rivers, covering each from its source to its end, from the Blackwater River in the western panhandle to the Miami River in the southern peninsula. (pb)

Marjory Stoneman Douglas: Voice of the River—An Autobiography with John Rothchild by Marjory Stoneman Douglas. This is the story of an influential life told in a unique and spirited voice. Marjory Stoneman Douglas, nationally known as the first lady of conservation and the woman who "saved" the Everglades, was the founder of Friends of the Everglades, a feminist, a fighter for racial justice, and always a writer. (pb)

Parrots of South Florida by Susan Allene Epps. If you've ever seen a parrot flying among the trees in your neighborhood, you're not alone. The species

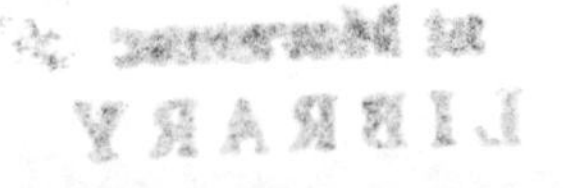

that inhabit Florida are detailed in this handy guide. You'll discover each bird's origin, description, size, habits, diet, and the towns and even specific neighborhoods where you can spot it. A stunning, color illustration by Karl Karalus accompanies each description. (pb)

Poisonous Plants and Animals of Florida and the Caribbean by David Nellis. Here's a list of the most menacing plants and animals around. Each is accompanied by a full-color photo to help you identify it. Then you'll learn what to do if you eat or come into contact with anything from poison ivy to a brown recluse spider. (pb)

Priceless Florida by Ellie Whitney, D. Bruce Means, and Anne Rudloe. An extensive guide (432 pages, 800 color photos) to the incomparable ecological riches of this unique region, presented in a way that will appeal to young and old, laypersons and scientists alike. Complete with maps, charts, and species lists. (hb, pb)

Stalking the Plumed Serpent by D. Bruce Means. Based on his forty-plus years of field research, the author reveals the biological complexity and exquisite beauty of the animals he has studied across the globe. Most people loathe these reptiles and amphibians, but Means loves and admires creatures that go bump in the night. (hb)

For children:

Iguana Invasion by Virginia Aronson and Allyn Szejko. How did nonnative animals, like seven-foot Nile monitor lizards and poop-tossing Rhesus monkeys, get into and begin to dominate Florida's ecosystem? Usually we humans are to blame. Find out how you can stay safe, appreciate exotic species, and treat them humanely. (pb)

Everglades: An Ecosystem Facing Choices and Challenges by Anne Ake. The fragile, unique ecosystem of Florida's Everglades is in trouble, and it's all about water. People have ditched, diked, diverted, and drained the water around the Glades for years. But if we destroy the Everglades, we destroy the water system of south Florida. Can we save the Glades? Can we undo the damage we've done? You decide. (hb)